# The Weird and ✦ Wonderful World of Bats

# The Weird and Wonderful World of Bats

Demystifying These Often-Misunderstood Creatures

## Alyson Brokaw

Timber Press

Portland, Oregon

Timber Press
Workman Publishing
Hachette Book Group, Inc.
1290 Avenue of the Americas
New York, New York 10104
timberpress.com

Timber Press is an imprint of Workman Publishing, a division of Hachette Book Group, Inc. The Timber Press name and logo are registered trademarks of Hachette Book Group, Inc.

Printed in China on responsibly sourced paper

Text and cover design by Leigh Thomas

The publisher is not responsible for websites (or their content) that are not owned by the publisher.

The Hachette Speakers Bureau provides a wide range of authors for speaking events. To find out more, go to hachettespeakersbureau.com or email hachettespeakers@hbgusa.com.

ISBN 978-1-64326-190-4
A catalog record for this book is available from the Library of Congress.

*In memory of Bill Brokaw, who taught me to love science and chase down my dreams.*

*To Pat Brokaw, for reminding me to stay curious and compassionate.*

*And to Mrs. Bat, the big brown bat(s) who started it all.*

# Contents ︿

# Introduction

The sun has finally dipped below the horizon, and the sky has softened from golden yellow and red to a more muted blue and purple. As the birds settle down for the night, a new chorus of sounds begins, with the croaky bass of frogs and the zippy vibration of crickets and katydids filling the night air.

Ears twitching, you give a great big yawn and stretch, one limb at a time. In doing so, you bump into your neighbor, who shuffles away with an angry squeak. Many velvety bodies ripple around you, swaying slightly as they shift from one foot to the other. You can just make out a soft glow in the distance, stark against the dark cave walls.

Wings slightly open at your sides, you take one last pause before dropping straight down. You fall, but as your wings finish unfurling you are suddenly buoyed back up. Swooping those wings forward and up, you orient toward the faint soft glow of the cave entrance. A nearby bat swoops by, brushing by the edge of your wing tips, though not enough to bump you off course. The air is filling with a cacophony of sliding chirps and buzzes as colony mates swirl around you. Riding the cooler currents of the outside air, you woosh out of the cave entrance. Around and around you circle, spiraling up and out into the slowly darkening sky.

▶ Mexican free-tailed bats fill the evening sky in central Texas.

# Welcome to the World of Bats

Bats have long captured the human imagination, woven into stories, mythology, and folklore in cultures across the world. Many stories paint bats as outsiders, due to their peculiar position as furred mammals with flight like birds. Their uncanny ability to navigate in the dark and preference for deep, dark places led to associations with the devil and witchcraft in western European cultures. This unfortunate alignment was further solidified into western popular culture by characters like William Shakespeare's witches in *Macbeth* ("eye of newt, toe of frog, wool of bat, and tongue of dog") and Bram Stoker's *Dracula*.

As a bat biologist, I'm often asked "Why bats?" I grew up among the fields and woods of eastern Pennsylvania. In the summers, my family and I would often eat dinner outside on the deck, where the gravity-defying antics of the local big brown bats could be spotted above the grassy fields. "Here comes Mrs. Bat," my dad would say, pointing out the dark silhouette of a bat as it twirled through the air above our heads. There is something truly mesmerizing about watching a bat in flight, an aerial ballet of movement that is so unlike any other flying animal.

As a high school student taking Advanced Placement biology, I was fortunate enough to be able to travel to St. John in the US Virgin Islands. One sunny morning, while exploring among the sugar plantation ruins along the Reef Bay Trail, I looked up to see several tiny bat faces clustered along the wooden beams of one of the stone buildings. Viewed through binoculars, this was my first time really *seeing* a bat, their expressive eyes, their fuzzy ears, and funny-shaped noses. Most significantly, these were not the small, black, scary creatures of legend. With over 1400 species currently described—and more counted every year—bats are the second most diverse group of mammals (rodents are first, with approximately 2500 species).

Diversity in bats is more than just the numbers, though. It's where and how these animals live. Bats can be encountered almost everywhere on the planet, from the Arctic Circle tundra to some of the most remote tropical islands. I often joke that if you can think of something unusual that a mammal might do, there is probably a bat species somewhere in the world that does it. So far, I've been stumped twice: there are no egg-laying bats (egg-laying mammals are very rare and are on a different side of the mammalian tree of life) and no bats rely

◄ The author holds a spectral bat, the western hemisphere's largest bat species.

on scavenging dead animals for food. And while it is unlikely that we'll ever discover an egg-laying bat, I wouldn't be all that surprised if future research uncovered evidence of bats as scavengers.

Even with continual advances in bat monitoring technology and genomic methods of analyzing species, diets, microbiomes, and everything in between, there is still so much to learn from and about bats. And for scientists, that's the best thing about bats. Just when we think we have them figured out, they go and surprise us. Since 2020, dozens of new bat species have been formally described. Some are cryptic species—species that can only be distinguished from each other by their genetic code or minute morphological differences like the overall shape of the skull or teeth. Others are completely new to science, like the Halloween-ready Nimba Mountain myotis. Captured in Guinea in 2019, this little orange-and-black bat was a complete surprise to the research team that made the discovery.

To enter the world of bats is like dropping into the multiverse. As an avid fantasy reader, I've dreamed of falling through a portal to another realm of magic and adventure. When I'm standing outside of a bat cave, feeling the swoosh of leathery wings against a darkening sky—those are the times where I start to feel like these other worlds just might exist. Where an echo is not just an echo, but a clue to the location of a tasty treat. Where the twitch of a finger can lead to amazing airborne acrobatics and the distant glow of the stars can guide you home.

By sharing the many fantastical ways that bats live and adapt to the world around them, I hope to bring these wonderful creatures out of the shadows and into the light. For as much as bats are unique, they are also like us in many ways. Bats argue with their roost mates and share with their friends. They push themselves long distances during migration to ensure enough resources for their growing young, and then eventually give those young the push they need to leave the roost and spread their wings.

And, importantly, the world needs bats. From insect control to pollination and seed dispersal, bats play critical roles in their ecosystems. They can help us better understand our own immune systems and might even hold the key to unlocking longevity. Come, step with me into the enigmatic world of bats.

# ORIGINS

## Chapter 1

Bats are mammals, so they share many important characteristics with humans. They give birth to live young, nourish the young via milk produced from mammary glands, have three little bones in the middle ear, and are furry and hairy.

Living bats are currently classified into twenty-one different families. Some bat families, like the Vespertilionidae (evening bats), are widely distributed across the globe. Others are found only in certain regions of the world. The New World leaf-nosed bats, a highly diverse group in the family *Phyllostomidae*, only inhabit the tropics and subtropics of North and South America. In contrast, flying foxes and other bats in the family Pteropodidae are found in many places *except* North and South America and most of Europe. Other families of bats are rarer still, living only in small pockets of the world. For example, the bumblebee bat, a tiny insect-eating bat with a wingspan reaching about 6 inches, has only been found in a few locations in Southeast Asia.

▶ The southern yellow bat has long been known as *Lasiurus ega*. Some scientists have proposed that yellow bats should instead be classified as *Dasypterus ega*, whereas others consider *Dasypterus* a subgenus of *Lasiurus*.

# Bats Around the World

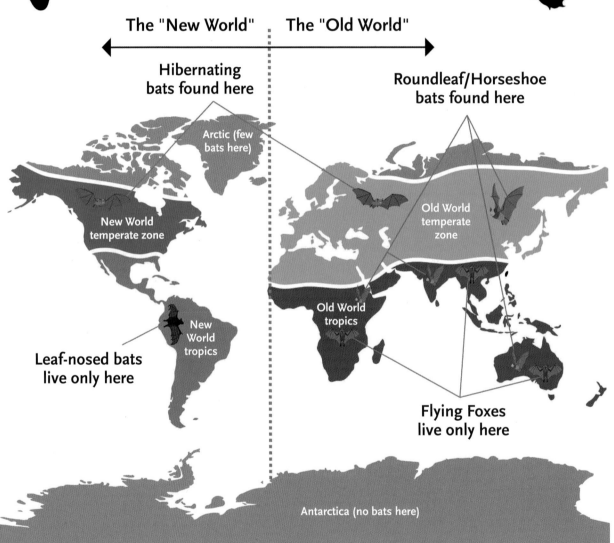

The "New World"  |  The "Old World"

Hibernating bats found here

Roundleaf/Horseshoe bats found here

Arctic (few bats here)

New World temperate zone

Old World temperate zone

New World tropics

Old World tropics

Leaf-nosed bats live only here

Flying Foxes live only here

Antarctica (no bats here)

While bats can be found almost everywhere on Earth, some types of bats only inhabit certain areas. The so-called Old World bats live in Africa, Europe, Asia, and Australia, and the New World bats inhabit North, Central, and South America. Hibernating bats are generally only found in temperate regions of the world, whereas fruit- and nectar-feeding bats are more limited to tropical and subtropical regions.

Within each taxonomic family, bats are classified further by genus and species. Together, these Latin binomials identify one specific type of animal; for example, scientists refer to the big brown bats of my youth as *Eptesicus fuscus*. Unfortunately, unlike in birds, common names for bats are not standardized, which can lead to similar and even overlapping common names depending on the region and local language. In addition to providing a unique name for each type of bat, binomials also reflect the evolutionary history of a bat and how it is related to other species. Bats that share the same genus (the first part of the scientific name) are assumed to be more closely related to each other than bats from different genera.

Researchers continue to uncover new information about these evolutionary relationships, however, which can trigger changes in the scientific name. And researchers don't always agree on the details of those evolutionary relationships, which can make things even more confusing. For example, the tree bats of North America are currently classified together in the genus *Lasiurus*. But recent studies have suggested that these bats are not as closely related to each other as scientists first thought and should be separated into three distinct genera: *Dasypterus* (yellow bats), *Aeorestes* (hoary bats), and *Lasiurus* (red bats). Others have argued that these groupings don't change the underlying evolutionary relationships and so instead consider those new names as subgenera. Depending on the author, it is possible to see the same bats classified slightly differently. For the most part, this scientific quibbling is insignificant and has little effect on what we know about the species themselves. If you want a basic list of current scientific names (or want to know the most up-to-date bat species count), check out the taxonomic and geographic database Bat Species of the World (BatNames.org). This database is managed by mammal curators and bat scientists from the American Museum of Natural History to keep track of confusing bat names and new species.

▲ A Cozumelan golden bat emerges from its roost. Its outstretched wings show the handlike shape that gives all bats their scientific name, Chiroptera or "hand-wing."

# Phylogeny of Bats

While the standard language of how scientists classify animal species has stayed consistent, the last century has seen some changes in how we classify and talk about bats. Scientists visualize evolutionary relationships with a phylogeny (or tree of life). The lines of the phylogeny indicate how different species have evolved through time. The tips of a phylogeny indicate extant species (or genera or families), the living organisms we see today. As you follow the lines away from these tips, places where they meet another line indicate a most recent common ancestor. Because we can never truly know what happened millions of years ago, these phylogenies are hypotheses or best guesses of how certain species may have evolved. As such, they are altered as science advances and we develop new techniques for understanding how genes and morphology change through time. The evolutionary history of bats is no exception.

All bats belong to the order Chiroptera (which roughly translates to "hand-wing"). Within the order Chiroptera, bats were historically divided into two major groups (or clades): Megachiroptera and Microchiroptera. Megachiroptera (or the megabats) included the large flying foxes and related fruit- and nectar-feeding bats that cannot echolocate, and Microchiroptera (the microbats) basically encompassed everything else. These early classifications were based mainly on behavioral traits like the presence or absence of echolocation and anatomical traits like inner ear morphology, jaw, and ear shapes.

It is widely accepted that all bats are descended from a single common ancestor, a kind of protobat that was scampering in the trees some 50 to 60 million years ago. In the early 1900s, mammalogists hypothesized that the closest relatives of these protobats and living bats were flying lemurs, primates, and treeshrews. However, starting in the late 1980s, advances in molecular tools for studying genetic relationships between organisms raised questions about these closest ancestors and even how bats are related to each other. The earliest studies based on nuclear and mitochondrial DNA sequences of bats revealed that it was unlikely that bats were that closely related to flying lemurs, primates, *or* treeshrews, proving those early hypotheses entirely wrong. Large-scale studies of mammalian genomes have further supported these game-changing studies. Based on the most recent genome studies, Chiroptera is now grouped under the superorder Laurasiatheria, along with hooved mammals, carnivores, and pangolins. In short, this means that bats are now understood to be more closely related to animals like deer, whales, and dogs. They truly are sky puppies!

Another curious pattern emerged from these molecular studies. Some bats formerly considered to be microbats (specifically the horseshoe bats) are actually more closely related to members of the Megachiroptera than to Microchiroptera. For a while, these ideas were attributed to analytical artifacts due to limited sampling and methodologies. In 2002, Emma Teeling and a team from University College Dublin set out to address this question once and for all, using nuclear DNA sequences from eleven bat families. Not only did they find that horseshoe bats grouped more closely with megabats, but so did the false vampire bats (Megadermatidae) and mouse-tailed bats (Rhinopomatidae). Based on these results, Teeling and her colleagues suggested a new classification for bats, replacing Megachiroptera and Microchiroptera with the suborders

▶ Mouse-tailed bats (genus *Rhinopoma*) are more closely related to flying foxes than they are to most other echolocating bats. Formerly considered a microbat, they are now classified under the suborder Yinpterochiroptera.

OPPOSITE:
▶ The current bat phylogeny. The points of connection represent where two evolutionary lines share a most recent common ancestor. Bats in the top seven families are believed to share a common ancestor and are grouped into the suborder Yinpterochiroptera. The rest of the bats are grouped as Yangochiroptera.

Yinpterochiroptera and Yangochiroptera. Within Yinpterochiroptera are the nonecholocating flying foxes and Old World fruit bats, plus the echolocating horseshoe bats, false vampire bats, roundleaf bats, and mouse-tailed bats. The rest of the bats are classified as Yangochiroptera.

## Studying Bats

Because bats fly and are nocturnal and generally small, most of us rarely encounter them in our daily lives. To study their lives and behaviors, scientists must seek out bats where they live. For many studies, this involves capturing and handling bats, using great care to minimize disturbance and harm to them. Mist nets are often used to capture bats as they are flying. As implied by the name, mist nets are soft, very fine pieces of netting that can be strung up over streams and ponds or along forest paths where bats might be flying. Bats are not able to easily detect the nets and therefore fly into them, where they hang until a scientist comes to gently untangle them. Near areas where many bats might be flying at once, such as outside a roost, scientists instead use a harp trap. This device has two or three rows of thin fishing line strung vertically, like a harp,

# The Bat Family Tree

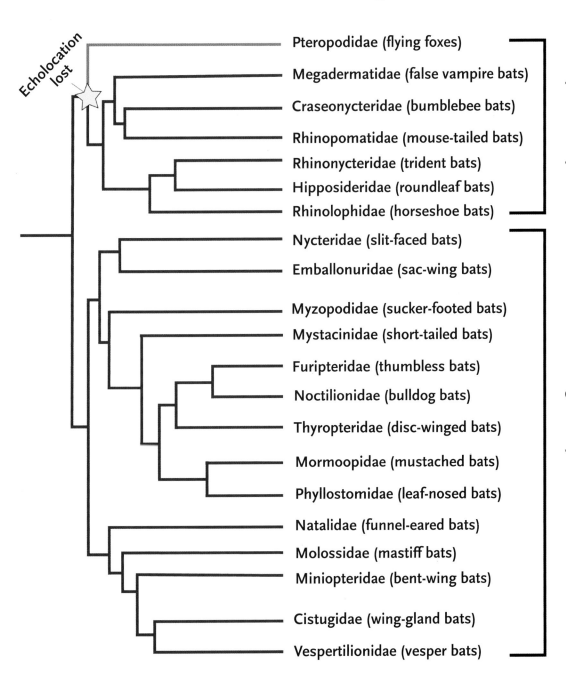

Echolocation lost

Pteropodidae (flying foxes)
Megadermatidae (false vampire bats)
Craseonycteridae (bumblebee bats)
Rhinopomatidae (mouse-tailed bats)
Rhinonycteridae (trident bats)
Hipposideridae (roundleaf bats)
Rhinolophidae (horseshoe bats)

Yinpterochiroptera

Nycteridae (slit-faced bats)
Emballonuridae (sac-wing bats)
Myzopodidae (sucker-footed bats)
Mystacinidae (short-tailed bats)
Furipteridae (thumbless bats)
Noctilionidae (bulldog bats)
Thyropteridae (disc-winged bats)
Mormoopidae (mustached bats)
Phyllostomidae (leaf-nosed bats)
Natalidae (funnel-eared bats)
Molossidae (mastiff bats)
Miniopteridae (bent-wing bats)
Cistugidae (wing-gland bats)
Vespertilionidae (vesper bats)

Yangochiroptera

▲ A silver-haired bat is caught in a mist net over a creek. While the netting looks obvious in this photo, it is difficult for bats to detect with echolocation and nearly impossible to see at night. Many a bat researcher (including the author) have accidentally walked into their own mist nets while working late at night!

◄ Flower bats wait to be retrieved from the soft canvas bag of a harp trap, set in front of a cave entrance in Jamaica.

and at the bottom of the lines is a soft bag. When bats hit the fishing line, they slide down into the bag, where they can comfortably hang in clusters and wait to be retrieved by the researchers.

After capture, researchers take careful measurements of the bats to help determine the species. Some species are easy to identify based on coloration or size, especially in areas with low bat diversity. Other species are more challenging, requiring measurements of features like ears, toes, or teeth, whereas some can only be reliably identified based on genetic samples. For research that requires tissue samples, such as population genetics or genomics to study the genetic source of bat adaptations, researchers take a small piece of wing tissue. This is done as quickly and carefully as possible to reduce discomfort to the bat and because the wing membrane heals quickly. During handling, other samples can be taken from bats, including fur clippings for isotope studies on diet and migration and blood samples for investigating immunity and disease.

Tracking individuals over time—whether to study how long bats might live or to figure out how they move around—requires some method of identifying individuals. One way is banding, where a small metal or plastic colored band is attached to the wrist of the bat, assigning each bat its own unique number. While still used in some studies, bands are only useful for identification if the researcher can get close enough to read the numbers. A newer method to mark individuals is using passive integrative transponder (PIT) tags. Like the microchips used in pet dogs and cats, these small devices are inserted under the skin, each with a unique identifying number. Captured bats can be scanned for their

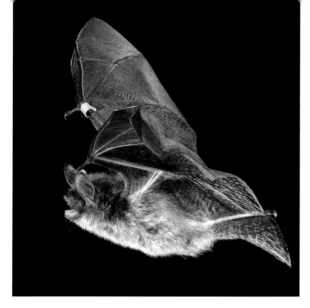

PIT tag number, but scanners can also be placed outside of roosts or areas where bats might fly through, allowing researchers to track bats remotely without having to capture the individuals again.

To track the fine-scale movements of bats over landscapes such as home ranges or areas where bats are foraging, researchers also use radio telemetry and global positioning system (GPS) units. To be attached to bats, these devices must be very small—preferably less than 3 percent of the bat's total weight. So, for a long time these units could only be used on the largest bats. Thankfully, advances in technology have not only made for smaller and smaller tags, but also tags that can track a bat's heart rate, body temperature, echolocation calls, speed, altitude, and more.

When emerging technologies are paired with the collaboration and knowledge of local peoples as well as the creative and inquisitive minds of scientists, the possibilities for new studies and discoveries are endless. While there is so much left to learn, let's take a peek into what's been uncovered thus far!

▼ This northern long-eared bat has a traditional metal band on its wrist.

▲ The author stands outside of a hibernacula cave entrance in northern Texas. The thick black cables hanging over the cave entrance are PIT tag readers. When a bat with a PIT tag flies close to those cables, its unique identifier will be automatically logged to a memory card, allowing researchers to track bat movements in and out of the cave.

# ECHO

Have you ever stood in an empty hallway and shouted, listening as the sound of your voice echoed around you? Even if you didn't realize it at the time, you'd just done an excellent bat impression! Now imagine being able to use those echoes to tell how far away the end of the hallway is, if a door is open or closed, or even what the texture of the walls is. One of the many remarkable things about bats is that they have adapted to use sound waves to detect and orient to the world around them, a process called echolocation. Also referred to as biosonar, echolocation is when an animal produces sound and uses the resulting echoes to collect information about its surroundings. In simplest terms, the animal does this by comparing what is said (the sound they produce) and what they hear (the echo). Bats are not the only animals known to use echolocation—others include some swifts, oilbirds, shrews, and dolphins—but bats have developed an amazing diversity and repertoire for exactly *how* they use echolocation.

▶ Do you emit echolocation calls out of your nose or mouth? Some bats, like this northern yellow-shouldered bat, might be able to do both.

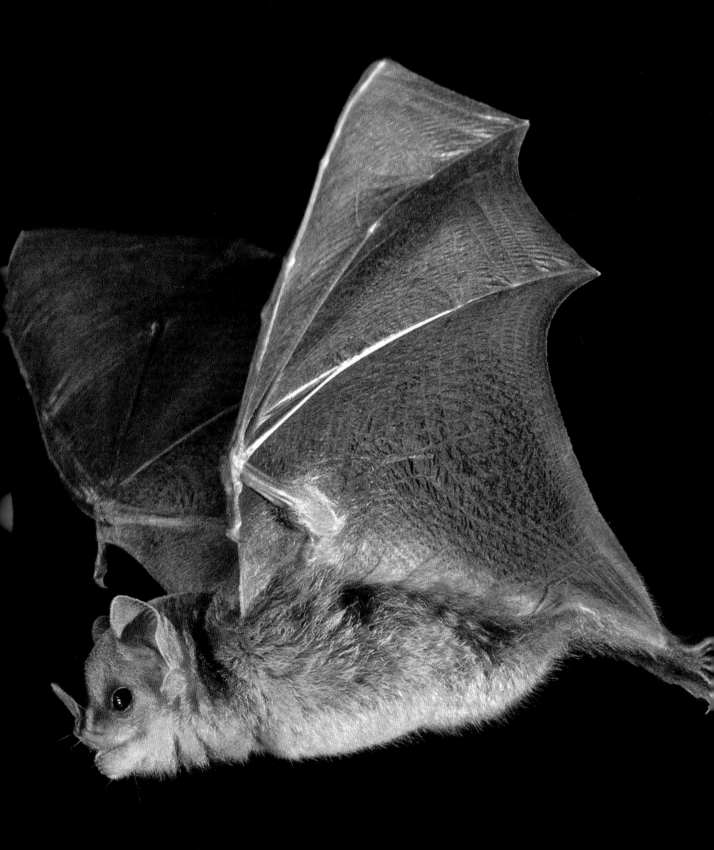

# Spallanzani's Bat Problem

Exactly how bats were able to navigate in the dark was a mystery for a long time. Lazzaro Spallanzani, an eighteenth-century Italian Catholic priest and naturalist, observed that, unlike nocturnal birds like owls, bats' abilities to navigate were not affected by loss of light. This curious observation led Spallanzani to conduct a series of flight experiments in 1793 in which he blocked the various senses of bats. He covered their eyes with blindfolds and plugged their noses with cotton, testing the roles of vision, hearing, smell, and even taste in bat navigation. None of these senses seemed to significantly affect bat flight, leading Spallanzani to suspect some sixth sense may be found only in bats.

A few years later, Swiss physician and naturalist Louis Jurine repeated Spallanzani's experiments, using long-eared bats and horseshoe bats. He also found that both species were able to navigate without trouble, even when blinded. Unlike Spallanzani, however, Jurine found that when bats' sense of hearing was impaired (via methods including wax plugs, tinder plugs, and even piercing the eardrum), bat flight became erratic or the animals refused to fly. Jurine concluded that "the eyes of the bat are not indispensably necessary to it finding its way . . . [but] the organ of hearing appears to supply that of sight."

These early scientists understood that hearing was important for bat flight, but exactly how sound was involved remained a mystery for more than a century—becoming known as Spallanzani's bat problem. How could bats use sound when they flew so silently? In the early 1900s, researchers took a fresh approach to the problem, considering that bats might detect obstacles by producing sound and listening for the reflecting echoes. The most plausible way bats could do this would be with high-frequency or ultrasonic sounds. There was just one problem: by definition, ultrasonic sounds are above the upper limit of human hearing.

George Pierce was a physicist at Harvard University, where he developed a salt-crystal oscillator that could transform ultrasonic noises into sounds audible to the human ear—essentially creating the world's first bat detector. A Harvard student at the time, Donald Griffin became interested in the theory that bats use high-frequency sounds to get around. In collaboration with Pierce and later Robert Galambos, Griffin listened in on caged and flying bats, confirming bats' use of ultrasound to avoid obstacles in flight and coining the term

▲ Italian biologist Lazzaro Spallanzani conducted some of the first documented experiments to investigate how bats get around in the dark.

*echolocation* in the process. In 1953, Griffin dragged a literal truckload of equipment to a pond near Ithaca, New York, and listened to bats as they hunted insects above the water. He noticed that foraging bats were adjusting their sound production patterns in flight. At this point, no one had suspected that bats could use echolocation to detect and capture small, fast-moving insects on the wing. Even Griffin himself was surprised, later writing that "echolocation of stationary obstacles had seemed remarkable enough, but our scientific imaginations had simply failed to consider . . . this other possibility with such far-reaching ramifications."

# Echo Echo Echo

Echolocation calls are small bursts of sound, produced by bats as they move through the environment. Essentially bats shout to see. Almost all echolocating bats produce calls using their larynx (or voice box) and so are referred to as laryngeal echolocators. In humans, bats, and other mammals, the larynx is made up of folds of tissues called the vocal cords. Air passes through these vocal cords, causing them to vibrate. How fast they vibrate depends on the tension of the folds, which are controlled by muscles within and alongside the larynx, and the speed in turn determines the pitch of a produced sound. The cricothyroid muscle is the main muscle responsible for lengthening and tensing vocal cords; in bats, this muscle is very large relative to their body size. A recent study found

Echolocation call emitted from bat

Returning sound waves

epiglottis

arytenoid
cricoid

vocal membrane
vocal fold

thyroid
cartilage

cricoid

dorsal side

cricothyroid muscle
cricothyroid membrane

ventral side

trachea

▲ Bats detect objects and obstacles around them by emitting echolocation calls and listening for the returning sound waves. Bats have special adaptations in their larynx (voice box), especially in the cricothyroid muscle, that enable the production of these high-frequency sounds.

that the cricothyroid muscle can achieve superfast contraction rates that help bats produce rapid echolocation clicks in flight.

Adaptations in the vocal cords of bats also help them produce low-frequency sounds as well. Some bats have a vocal range covering almost seven full octaves—even the most accomplished human singers can only cover about three or four octaves. To produce low-frequency calls for communication, bats vibrate a set of false vocal cords (also called ventricular cords) that lay on top of the true vocal folds. Interestingly, humans can also vibrate these false vocal folds, which is how performers like throat singers and death metal vocalists produce those guttural, vibrating sounds.

When echolocating, a bat compares the time delay between the produced noise (echolocation pulse) and the returning echoes. The durations and between-call intervals (how much time elapses between each call) are usually measured in milliseconds and vary depending on the bat species, the task the bat is trying to accomplish, and the surrounding environment. The frequency (or pitch) of these calls also varies across species. Most bats produced echolocation calls above the range of human hearing (above 20 kHz), but some, like the spotted bat, produce calls much lower in pitch—as low as 8 kHz. If you live in an area with these low-frequency bats (such as the American Southwest for spotted bats), listen closely at night for the sound of birdlike, short, sharp chirps. What you might be hearing is a bat! On the other end of the spectrum, the trident bat produces calls reaching over 200 kHz.

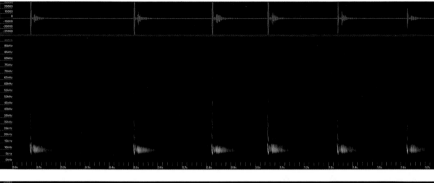

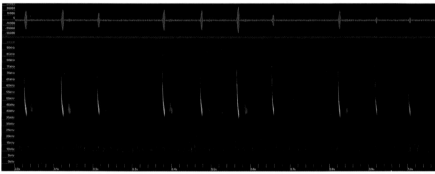

Most insect-eating bats produce echolocation calls between 20 and 60 kHz. Why such high frequencies? High-frequency wavelengths and their associated echoes provide better detail than low-frequency wavelengths. It's the acoustic equivalent of using a small, fine paintbrush to color in detail in a painting instead of a large roller.

# The Shape of Sound

The echolocation calls of a flying bat can be categorized into three phases: search, approach, and terminal buzz. During the search phase, bats produce less-frequent echolocation calls, with longer between-call intervals. These types of calls are often associated with bats flying in open spaces, where there are minimal obstacles to avoid. After detecting the echo from a target, such as a flying insect, the bat enters the approach phase, when both the call rate and the length of each call increase. As the bat comes closer to its target—as judged by decreasing time between echoes—it enters the terminal buzz phase. The bat

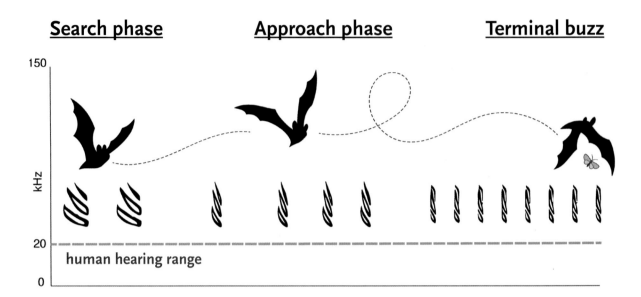

## Search phase     Approach phase     Terminal buzz

150

kHz

20

human hearing range

0

▲ The sequence of echolocation calls as a bat nears its target, a tasty insect. Search phase calls are produced at slow rates. Once an insect is detected, the bat increases the call rate before using the terminal buzz to zero in for the final attack.

starts emitting echolocation pulses very quickly, reaching maximum rates as high as 200 calls per second. These rapid pulses are also sometimes called feeding buzzes, because they usually indicate a bat has attacked or attempted to attack an insect. By varying the echolocation call rate and intensity, the bat is working to maximize the amount of information it obtains from the acoustic environment. While not exactly analogous, it's a little bit like squinting at a picture to resolve more detail.

One truly remarkable thing about bat echolocation is the overall diversity in echolocation signals. Different species of bats produce different echolocation calls, which can tell us about what and how a bat eats and where it lives. Bats that fly in open spaces, such as above the forest canopy or in open fields, tend to have longer, lower frequency calls that allow them to detect prey from farther away (up to 10 m). In contrast, bats that forage inside or on the edges of forests produce shorter calls in varying shapes to avoid confusing prey echoes with that of surrounding vegetation. These differences in echolocation calls represent a type of ecological partitioning, where species specialize on certain prey or habitats to avoid competition with other bat species in the area.

Each echolocation pulse produced can cover a range of different frequencies. Most bat calls are classified as frequency-modulated (FM), meaning the call starts at a higher frequency and then swoops downward to end at a lower frequency. These calls help bats navigate a variety of habitats and catch a range of

◄ Mesoamerican mustached bats living in the forests of Central America navigate using constant-frequency echolocation calls.

different types of prey. Other bats have evolved calls specially adapted for foraging in highly cluttered environments. These bats make constant-frequency (CF) calls that are dominated by a single frequency. Bats that produce CF calls include Old World leaf-nosed bats (Hipposideridae), horseshoe bats (Rhinolophidae), and some mustached bats (Mormoopidae). When visualized, CF calls look a little like a half-drawn table: a straight line with a small little peg (FM sweep) at the beginning or the end. Bats that make these calls are also called flutter-detecting bats, for their ability to recognize the flutter of insect wings.

Doppler-shift compensation is a trick used by bats that produce CF calls. As the source of a sound moves, the frequency of the sound waves changes relative to the observer. This phenomenon is called the Doppler shift and it's why ambulance sirens or train whistles become higher pitched as they approach us and then change to lower pitches as they move away from us. In this case, the bat is both the source and observer of a sound. As a CF bat approaches an insect, it adjusts its calls to account for its speed and listens for the echoes at a slightly lower frequency—the Doppler-shifted frequency. Imagine having to look for a

▲ Trefoil horseshoe bat emits constant-frequency echolocation calls from the nose.

green object against a background of many different colors. If your eyes were particularly sensitive to green and less sensitive to other colors, it would help that green object stand out. Bats that use Doppler-shift compensation can tune their auditory system to a very focused set of frequencies, which helps them pick out small details from other, potentially confusing information.

Scientists visualize bat sounds by plotting time (usually in seconds or milliseconds) against the pitch or frequency of the call. This sound picture (or sonogram) allows scientists to analyze different aspects of a bat's call, such as overall shape (CF versus FM), the duration of different types of calls, or how long the bat waits before calling again. Eavesdropping on the echolocation calls of bats can give a lot of useful information about what bats are around and what they are doing in their natural environment. Since different species produce different calls, it's possible to identify which species are in an area by looking at these calls, either visually or with software programs designed to discriminate between bat calls.

Whether you are a scientist collecting data or an enthusiast just interested in the bats in your backyard, listening in on the sounds of bats still requires specialized equipment. Luckily, we have come a long way since Donald Griffin's salt-crystal oscillator and truckful of listening equipment. Over the years, bat detectors and the associated equipment to record and visualize bat calls have gotten both more sensitive and compact. There are also a few different types of detectors that let us listen in on bats, but all bat detectors require special microphones that are sensitive enough to pick up on high-frequency bat calls.

Heterodyne detectors work by filtering for a certain sound frequency, then mixing it and shifting it downward into a frequency humans can hear. Because these devices are set to detect a certain frequency, they must be manually tuned and require some knowledge of the types of bats that might be in the area. Another method used by bat detectors is frequency division, in which the frequency (pitch) of the signal is divided in a way that results in a lower pitch. Finally, time expansion works by slowing down the call, which results in the pitch shifting downward. Unlike heterodyne and frequency-division detectors, however, time expansion is not useful for real-time listening to bats.

In the last decade, advances in computer processing and storage have made it easier and cheaper to directly record the sounds of bats by using full-spectrum or direct-sampling detectors. With these detectors, sounds are recorded at their original frequency, limited only by the device's sampling rate (number of sound samples per second). One benefit of these detectors is that they can record audio files for later analysis and, in some cases, also provide real-time monitoring. Some companies have even started making full-spectrum bat detectors that can be plugged directly into smartphones, allowing the user to watch and identify bat calls in real time. These types of detectors are best for activities like educational bat walks or transect surveys, where a scientist walks a set path to listen for bats in the area.

For long-term monitoring, a variety of passive bat detectors can be attached to a tree or pole and left in the environment for a while. These tough, waterproof devices are set to trigger when they detect a bat call and then record the audio file to a memory card, providing a way for scientists to identify which bats are using a particular area or habitat over the course of days or weeks.

▲ Waterproof bat detectors like this one can be set out at field sites for days or even weeks, recording any nearby bat activity.

▶ The flattened tragus can be seen poking up from the center of the ear of this hairy legged myotis, a small insect-eating bat that lives in Central America.

# What Big Ears You Have

The mammalian external ear consists of the pinna, concha, tragus, and external auditory meatus. The pinna is the fleshy part that sticks up or out from the head. Some echolocating bats have pinna with elaborate folds or exaggerated sizes compared to those of other mammals. Just like cats and dogs tilt or twitch their ears in response to sound, bats are also able to move parts of their ears to maximize detection of faint returning echoes. Bats that hunt primarily by listening to prey-generated sounds (like the rustling of leaves) tend to have extra-large ears. For example, the spotted bat has ears that when fully extended measure almost 2 inches, almost half of their entire body length. Bat ears also have a tragus, an extra fleshy protrusion that sits toward the front of the pinna. As sound enters the ear, the tragus creates a second echo as the sound reflects around the inside of the ear. Bats use the delay in this secondary echo to figure out where a sound came from in vertical space, whether above or below them.

After a sound has reached the outer ear, it travels down the external auditory meatus to the tympanic membrane, or eardrum. Small bones called ossicles connect the eardrum to the inner ear, helping transmit sound waves through the middle ear to the cochlea, a snail-shaped structure of the inner ear that is lined with sensory hair cells and filled with fluid. As sound enters the cochlea, it vibrates the hair cells, triggering nerve impulses that get sent down neurons to the brain via a bony tube called the Rosenthal's canal. The shape of the Rosenthal's canal differs among bats that produce different types of echolocation calls. Bats producing CF calls have the typical solid bony tube, whereas those that produce varying FM calls have large holes in the wall of this tunnel. Bats with these more open canals also have more neurons in this region, reaching as many as 46,000 neurons compared to the 17,000 found in CF bats.

# Bat Sounds in the Brain

From the cochlea, these many neurons travel to the auditory cortex, the main brain structure responsible for processing acoustic information. Different sub-regions of the auditory cortex respond to different elements of sound, such as frequency or, in the case of bats, the delay between emitted echolocation pulse and the returning echo.

Individual neurons in the auditory cortex respond to certain sound frequencies and are distributed in what is called a tonotopic map. In most bats, the front part of the auditory cortex has neurons that respond to high-frequency sounds, with lower frequency sounds triggering neurons toward the tail end of the cortex. Bats also have specialized delay-tuned neurons that allow them to precisely estimate the distance of an object from its echo. Bats estimate their velocity and the speed of the object of interest in another part of the auditory cortex called the CF-CF, named for its involvement in processing constant-frequency sounds. Finally, bats that use Doppler-shift compensation to detect moving prey have a third part of the auditory cortex that responds to changes in echo frequency as an object moves relative to the bat.

Bats can process acoustic information from echoes with impressive speed and accuracy. A recent study found that Mexican free-tailed bats can accurately

► Mexican free-tailed bats can accurately discriminate texture using echolocation alone.

discriminate between different textures. Bats were trained to associate coarser grit sandpaper with a mealworm reward. They easily discriminated between ten different sandpaper comparisons, struggling only when the difference in mean particle size (coarseness) was about 20 μm. Put another way, these bats were able to differentiate texture using sound with about the same accuracy of a human hand discriminating texture using touch.

## The Better to Hear You With

A downside of high-frequency calls is that sound doesn't travel as far as low-frequency sounds. How do bats compensate for the lost distance? By being really, really loud. Sound intensity is measured in decibels (dB), representing the sound pressure level at a specific distance away from a source. Leaves rustling in the breeze is about 30 dB, average human conversation tends to be about 60 dB, and running the vacuum cleaner is about 75 dB. The average intensity of the echolocation calls of a big brown bat flying in the open sky is around 130 dB—the equivalent of being front row at a rock concert or in the pit at a stock car race. Other bat species may call even more loudly, with the Neotropical lesser bulldog bat recorded calling at intensities greater than 140 dB.

On the flip side, so-called whispering bats generally produce softer calls, usually between 50 and 90 dB. Like humans, bats have control over how loud they are, and even these quiet bats have been recorded calling at intensities above 100 dB.

Variation in loudness is linked to what bats eat. Bats that capture insects out of the open air tend to be louder, although there are disadvantages for being too loud. Those bats that specialize on moths that can hear ultrasonic calls have adapted to produce very quiet calls to avoid detection by their prey.

If bats are producing echolocation calls at the intensity of a rock concert, how do they manage to still hear the returning echoes? Luckily, bats seem to have found a solution, called forward masking or self-deafening. When a bat calls, a small muscle in the ear called the stapedius contracts. This contraction reduces the movement of the ossicles and the amount of sound that is transmitted to the cochlea. All of this happens very quickly, allowing the bat to still hear faint echoes as they return to the ear following an echolocation pulse.

## Weird Noses

Four groups of bats have fleshy flaps and projections around their nostrils, called nose leaves: horseshoe bats, named for the horseshoe shape of their noses; roundleaf bats, with large, dish-shaped projections around their noses; false vampire bats, with long, flattened or heart-shaped nose-leaves; and the New World leaf-nosed bats, with nose leaves in all shapes and sizes ranging from large pointy spears to barely any leaf at all. While geographically distributed in different areas arounds the world, these flappy-nosed bats all have one major thing in common—they echolocate out of their nose.

The folds and flaps around the noses of these bats have evolved to improve the shape and direction of their echolocation calls. In the same way that we cup our hands around our mouths to better direct a shout, the projections above the nose act as baffles to direct the call. As the bat echolocates, the outer edges of the nose leaf twitch inward or downward, often extremely rapidly. The exact way these twitches shape the sonar beam varies depending on the species and the characteristics of the call (such as FM or CF calls), but in many species they serve to focus or narrow the beam in one direction. Horseshoe bats also have enlarged nasal chambers and unique strand-shaped nasal bones that seem to help them generate calls from their nose.

The conventional wisdom on bat echolocation calls was if a bat had a nose leaf, it only echolocated out of its nose. In many species of New World leaf-nosed bats, this presumably meant being able to echolocate while also carrying a large

  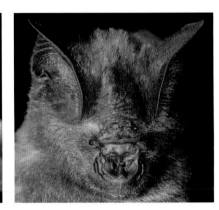

▲ The hairy big-eyed bat is a New World leaf-nosed bat.

◀ The rufous horseshoe bat, an Old World leaf-nosed bat, has a rounded and folded nose leaf.

▼ Schneider's leaf-nosed bat, an Old World bat, echolocates from its nose.

OPPOSITE:
▼ The Jamaican fruit-eating bat flies with its mouth closed. This species is never seen flying with its mouth open and presumably uses only nasal echolocation.

▶ While capable of echolocating with the nose, the northern yellow-shouldered bat—seen here flying with its mouth open—may also echolocate with its mouth.

piece of fruit or prey item in your mouth (got to love a multitasker). However, time and time again, several of the assumed nose-shouters were photographed flying with their mouths wide open.

To address the question of whether having a nose leaf means bats only echolocate out of their nose, a group of researchers from the Smithsonian Tropical Research Institute in Panama and the University of Ulm in Germany captured forty species of leaf-nosed bats in Panama and Peru. Led by graduate student Gloria Gessinger, the researchers flew the bats in an outdoor flight cage and recorded their behavior using high-speed video cameras and ultrasonic detectors. What they found was a little bit surprising. Some of the leaf-nosed bats, like the Jamaican fruit-eating bat and the hairy big-eyed bat, only flew with their mouths closed, consistent with the hypothesis that they rely solely on nasal echolocation. However, many of the species with a large and distinctive nose leaf, like the yellow-shouldered bat and short-tailed fruit bat, consistently flew with their mouth open. Other bat species seemed to use a combination of both strategies, sometimes flying with the mouth open and sometimes closed. While primarily observational, this study raises interesting questions for how, why, and when bats might echolocate from the nose versus from the mouth. It also highlights the importance of challenging assumptions when it comes to bats— and science!

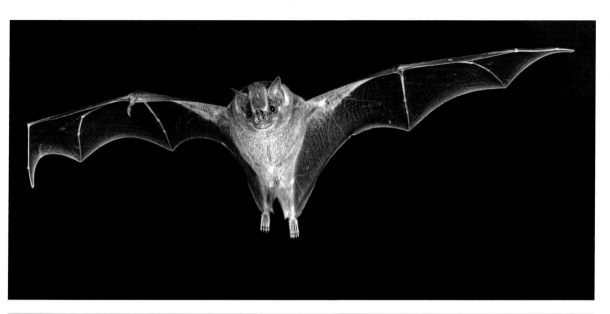

# Tongue Clicking

The use of ultrasonic pulses to navigate in the dark is an evolutionary innovation that was important to the evolution and diversification of bats across the world, but not all bats can use echolocation. Flying foxes and other Old World fruit bats in the family Pteropodidae don't echolocate, instead relying on vision and smell, with one interesting exception: the Egyptian fruit bat.

Egyptian fruit bats echolocate, but they don't do it like the rest of the world's echolocating bats. Instead of producing echolocation calls from their larynx, they click their tongues. Lingual echolocation and the associated clicking signals are considered less sophisticated than the signals produced by laryngeal echolocators, but that doesn't necessarily mean they are less effective. Tongue clicks produced by Egyptian fruit bats can be as short as 50–100 microseconds and provide enough information to help them avoid wires and other obstacles in the dark. The bats don't just emit one click at a time, instead producing clicking signals in pairs, separated by about 20 milliseconds. A study using a combination of high-speed cameras and microphones arrayed to track bats' head movements while echolocating revealed Egyptian fruit bats can adjust sonar beam directions without moving their head or lips—something it was thought only laryngeally echolocating bats could do. The trick seems to be in the location of the tongue during clicking. Using models, the researchers showed that a 6-mm shift in the tongue position could cause almost a 26° shift in beam direction during echolocation.

So why do Egyptian fruit bats use lingual echolocation, whereas other Old World fruit bats don't? The main hypothesis is that it has to do with their roosting ecology. Unlike other fruit bats that spend their days among the leaves and branches of trees, Egyptian fruit bats roost in colonies deep in caves, where there is little light available for vision. Using their tongue clicks, the bats can use sound to avoid running into cave walls while exiting the roost. Once in the open skies, Egyptian fruit bats can then also use vision and olfaction to get around.

Even with adaptations for low-light vision (more on that later), navigating in the dark without echolocation is still a challenge. And it turns out that at least some other Old World fruit bats have another trick up their sleeve (er, their wing). At least three species of fruit bats have been recorded producing clicking noises with their wings while flying. To test if these wing clicks were being

◄ The only member of the Pteropodidae known to echo-locate, Egyptian fruit bats produce calls with the tongue instead of the larynx.

used by the bats to navigate, researchers flew them in dark and light conditions, challenging the bats to navigate around objects like thick wire. Untrained bats increased the rate of clicking when flying in the dark, suggesting the use of the sounds for improving navigation in low-light conditions. Wing clicking does not appear to give bats a lot of detail though, as they generally failed to completely avoid the wires in the dark.

Exactly how these bats produce clicks with their wings remains a bit of mystery. Even after using high-speed cameras to look at the bats flying in dark and light conditions, there were no obvious differences in the ways their wings moved. It's also unclear just how widespread this behavior might be in these fruit bats and just how much the animals rely on these signals. Compared to lingual and laryngeal echolocation signals, wing clicks are much quieter and lower in frequency, making it difficult to detect them without highly sensitive equipment. Even if these signals are only occasionally used by these bats, they might still offer interesting insights into how bats evolved to use echolocation. This discovery is also a reminder of the many ways that bats can continue to surprise us.

# SCENT

## Chapter 3

When I was living in Gamboa, Panama, as a research fellow at the Smithsonian Tropical Research Institution, I walked by a large fig tree every day on my way to the lab. The tree was nestled in the side of a hill, with branches that hung low over the sidewalk, requiring me to duck or walk around them to avoid getting snagged. This was in early September, solidly in the middle of the Panamanian wet season, which is generally warm and muggy. I didn't take much notice of the tree in the first few weeks I was in Gamboa. Meanwhile the fig tree was just being a regular tree, doing normal tree things—soaking in carbon dioxide, turning it into sugar, cleansing the air we breathe.

▶ Even in bats that echolocate, smell is an important sense. Fruit eaters like this great fruit-eating bat use smell to help them find and make decisions about food.

After a few days taking a different route to the lab, I noticed something different about the tree. Many fig tree species in Central and South America undergo a big bang process of fruiting, where all the fruit on a single tree ripen at the same time. The dramatic explosion in ripeness attracts a whole range of fruit-eating animals, including monkeys, toucans, parrots, agouti, kinkajous, and bats. When the neighborhood tree began to ripen, it dropped little ripe green figs all over the street and occasionally on your head (if you weren't careful). Beyond having to dodge the potential bonk to the noggin, the fruit stunk. The smell was probably due to the overripe figs that had dropped to the ground and got trampled into a slick, stinky, slip-and-slide of squashed fruit. When I curiously plucked a green fig and held it up to my nose, however, it did have a fruity, but faint smell.

We humans and many other primates rely primarily on vision to sense and navigate the world around us. Plants that want to attract primates and birds as seed dispersers tend to produce fruits that change colors as they ripen, from an inconspicuous green while growing to a bright red or purple at ripeness. In contrast, bat-dispersed fruits don't change color as they ripen, generally staying a mottled green color, so the figs don't stand out very much against the bright green leaves. So, how do bats know when ripe fruits are available? They can smell it!

An important characteristic of bat-dispersed figs and other fruits is their fruity smell. The exact fruit scent varies by plant species, and the scent profiles of the fruit change both with stage of ripeness and time of day. While most research on the sensory world of bats has focused on their amazing ability to use sound to navigate the world, only recently has there been increased focus on bats' other senses like smell. So, what's going on inside a bat's nose?

## Do You Smell What I Smell?

Traditionally, animals were separated into two distinct categories based on how well humans thought they could smell. Sniffing savants, like the domestic dog or the house mouse, were called macrosomats, animals with a keen sense of smell. On the other side were microsomats, animals assumed to have a poor sense of smell. These less-than-stellar sniffers included humans and bats. But as we've learned more about how animals smell, these classifications might not be as distinct as we thought. Does being a macrosomat mean an animal detects many different smells? Or does it mean the animal is highly sensitive to certain smells but not others?

Olfactory physiology is complicated enough that Linda Buck and Richard Axel won the Nobel Prize for Medicine in 1991 for first describing how olfactory receptors and their associated genes work to tell our brains about smell. When a mammal smells, it pulls airborne odor molecules over a special layer of cells in the nose called the olfactory epithelium. Embedded in the olfactory epithelium are a bunch of different olfactory receptors. Each of these receptors has a different shape and binds to a particular odor molecule, which sends impulses down the olfactory sensory neurons—special cells that send electric impulses to the brain. The neurons deliver information about these bound molecules to the olfactory bulb at the front of the brain, where things like identity and odor concentration get sorted out. In theory, the greater variety of olfactory receptors an animal has, the more types of odors it can tell apart (discrimination). Similarly, the more olfactory receptors an animal has of a certain type, the more likely that animal will be able to detect a certain odor, even if only a few molecules are floating around (sensitivity). This makes certain animals highly sensitive to certain odors and able to track them with incredible accuracy.

Olfactory receptors are coded by olfactory receptor genes. Collectively, these genes form some of the largest gene families in vertebrates, and the various

▼ Fruit- and nectar-feeding bats rely on smell to help them forage. Odors floating in the environment are breathed in through the nose, where they bind to olfactory receptors in the olfactory epithelium. Once the odor molecules are bound, neurons send the information to the brain to help the bat identify, discriminate, and find food.

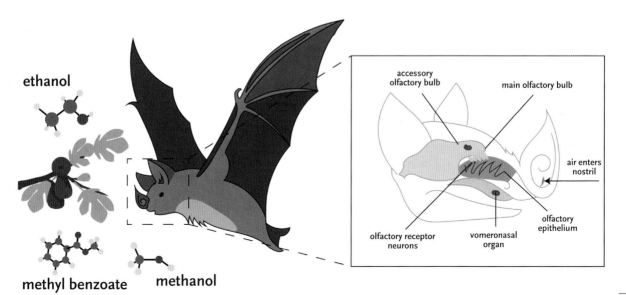

ethanol

methyl benzoate    methanol

accessory olfactory bulb

main olfactory bulb

air enters nostril

olfactory epithelium

vomeronasal organ

olfactory receptor neurons

olfactory receptor gene families detect and respond to different sets of odors (though there is some overlap). Different vertebrate species have diverse collections of these genes. Among mammals, African elephants have the largest known collection of olfactory receptor genes, numbering almost 2000 compared to the 800 or so in dogs and 400 in humans. The habitat and diet of a species play important roles in how these olfactory receptor gene repertoires evolve. Ancient bats are thought to have undergone a loss of olfactory receptor genes, evident today in most echolocating bats. This gene loss is not as pronounced in Old World fruit-eating bats, which don't echolocate, suggesting that contraction may be related to a trade-off between echolocation and olfaction. Both echolocating and non-echolocating fruit bats also have increased variation in two specific groups of olfactory receptor genes, OR 1/3/7 and OR 2/13, which may be linked to the evolution of fruit eating in bats.

It takes more than just good genes to be a good smeller. It also takes the right nose and brain anatomy. The more olfactory sensory neurons that connect to the brain and the bigger the olfactory bulb (relative to total brain size), the more likely it is that an animal relies heavily on smell. Fruit-eating bats tend to have relatively larger olfactory bulbs than those of insect-eating bats, reflecting a greater use of odor cues when foraging. Some mammals known for their

► Bats that consume fruit, like this Jamaican fruit-eating bat, have larger olfactory bulbs and more olfactory surface area than insect-eating bats.

◄ Insect-eating bats, like this little brown bat, have a weaker sense of smell than fruit-eating bats.

sense of smell, like dogs and cats, also have an intricate labyrinth of thin, scroll-shaped bones inside their nose. Called ethmoturbinals, these bones are covered in olfactory epithelium and help maximize the amount of surface area for detecting odors in the air. While not as intricate as a dog's, Jamaican fruit-eating bats have more intricate ethmoturbinals and nearly twice as much olfactory surface area as insectivorous little brown bats.

## Tracking Down Fruit

When I was at the Smithsonian Tropical Research Institution, about an hour before sunset I would head into the lab to prepare for the night's experiments. After checking on my Jamaican fruit-eating bats in temporary captivity, I'd head to the flight room to turn on infrared lights and the two cameras, each one tucked into a corner. The flight room was about a 16-foot-by-16-foot square, with a cement floor and dark black curtains covering its metal mesh walls. Along the far side of the room, I placed a row of wooden platforms, each about 4 feet high. While I prepped the flight room, my assistant Evynn painstakingly cut small

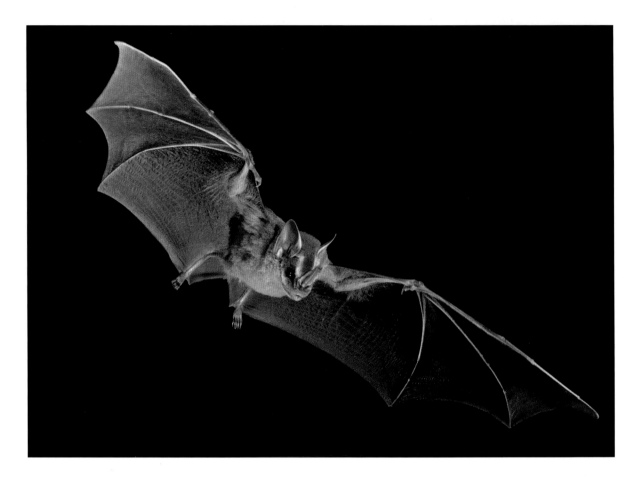

▲ Small fruit-eating bats can discriminate ripe and unripe fruits when foraging.

pieces of ripe banana and placed them in little plastic weigh boats. The goal of this project was to understand how bats might use odors—specifically if bats can track odors right to the source.

Thirty years ago, on Barro Colorado Island (only 16 km from where I was working in Gamboa), Elisabeth Kalko and her team conducted the experiments that laid the foundation for what we understand about bats, fruits, and fruit smells. They were interested in exploring the dynamics between echolocation and olfaction in the small, leaf-nosed, fruit-eating bats common to the island (*Carollia*, *Artibeus*, and *Vampyressa*). While echolocation is a great adaptation for navigating and discriminating objects in the dark, the clutter of leaves and branches commonly found in the thick rainforest understory can create confusing echoes and make it difficult for bats to find a particular branch or hanging fruit.

In a series of experiments, Kalko and her colleagues offered bats ripe, unripe, and scented artificial fruits. They found that bats consistently chose the fruits that smelled right—like ripe, tasty fruit. In later experiments, they even mashed up ripe and unripe fruits and wrapped them in cheesecloth, creating indistinct balls of gooey fruit. Even when the fruits were unrecognizable by shape, bats were still attracted to the smell of ripe fruit, ignoring unripe or fake odors. This extra step provided good evidence that bats are using odors to choose fruits and not echolocation or vision. These and similar experiments also suggested that odors might be an important signal to help bats find fruit over long distances.

The ability to follow odor plumes is not uncommon in the animal world, but it is more challenging than one might think. Imagine spraying a small amount of perfume in one corner of a room. If you're standing right next to the spot the smell will seem very strong, but on the other side of the room it might not be detectable at all. That is because the number of odor molecules (concentration) is highest at the initial source but decreases as the molecules move away from that source in a process known as diffusion. Eventually, enough of those odor molecules will make their way to the other side of the room to be detected, traveling via diffusion and air movement.

At small scales, animals can follow that simple gradient, using a strategy known as klinotaxis: moving their body or head in the direction of the stronger odor concentration. Animals can also estimate the direction of an odor (and really, any kind of sensory signal) by comparing the strength of signal from different receptors. In mammals, that means comparing the strength of a smell between the right and left nostrils. Unfortunately, air movement breaks up this nice straightforward gradient, creating swirling odor plumes with pockets of relatively odorless space. Flying animals like moths and other insects are adept at tracking these churning plumes, moving in zigzag patterns across the plumes to detect the odor before moving forward. Dogs and other mammals perform similar behaviors when tracking trails of odors on the ground. Given their airborne habits, I wondered whether bats could do this too.

Back at the flight room, once the sky was fully dark, it was time for lights, camera, action. One fruit-eating bat was released into the flight room at a time, illuminated only by infrared light (light waves not visible to humans or bats). On a computer screen just on the other side of the wall, Evynn and I watched the bat zip around the room in wide circles, before it eventually approached and landed on one of the wooden platforms. After a moment of awkward scrambling, the bat successfully grabbed a piece of banana before flying off to munch its prize. A quick break to capture the bat and reset the odor sources and we'd repeat the test with a new bat. In addition to actual banana, bats were sometimes presented with only the smell of banana to see if they could find the source of the delicious smell. (Bonus: it made the lab smell like banana bread!)

Using the videos and specialized software, I then recreated the flight path of the bat as it searched the room for the tasty smells. If the bat was able to follow the path of an odor plume, we expected to see the zigzag casting behavior as it approached the scent platform. Instead, the bats spent time circling the whole flight room before swooping down to check out one or more of the platforms. While they eventually landed on the spot containing the scented reward, it seemed they still needed to get close to the scent before deciding to land.

Our findings suggest that, although smell is still important for making decisions of where and what to eat, it may not play a huge role in helping bats find the right tree. Instead, bats might use mental maps of an area to remember where trees are, then use scent to pick only the tastiest food.

## Smells Like Fruit Spirits

Without color cues to know when fruit is ripe for the picking, fruit-eating bats around the world rely on smell to separate the good from the bad. Of course, bats aren't the only mammals who use smell to make decisions about food. If you've ever let a piece of fruit ripen too long, then you know the musty, slightly alcoholic smell that accumulates. After the fruit has reached the peak of ripe deliciousness, the sugar begins to be converted into alcohol, mainly ethanol. This conversion is carried out by bacteria and yeast in a process called fermentation. It's the same wonderful chemical reaction that gives us libations like beer and wine. Alcohols, like ethanol and methanol, are highly volatile compounds, meaning they easily transition into a gas from their liquid form. As they enter the gaseous phase, the molecules move through the air and become detectable by smell. When fruits ripen and begin to ferment, the amount of these volatile compounds released increases, providing a way to evaluate the ripeness and palatability of fruit. This ability to use smell to evaluate food is common across

many animals and is one of the main ways we, as humans, avoid ingesting potentially dangerous underripe or spoiled foods.

Ethanol levels in naturally ripe fruit vary, ranging between about 0.1 to 4.5 percent ethanol. There's a certain alcohol level that signals the Goldilocks spot for fruit—not too ripe and potentially dangerous, but not too underripe and hard to digest. When given a choice, fruit bats overwhelmingly select the smell of fruits in their "just right" category. Methanol and ethanol are some of the scent cues fruit-eating bats use when foraging. Scientists originally thought that bats might be strongly attracted to ethanol, using the smell to locate fruit over long distances. When researchers presented Egyptian fruit bats with mango juice spiked with varying levels of ethanol and methanol (0.01–2 percent), the bats showed no preference for slightly spiked cocktails (0.7 percent or less) compared to plain mango juice. However, they did strongly avoid more alcoholic (above 1 percent) choices.

There are two main reasons why these bats would avoid fruits with high levels of ethanol. First, as in humans, ethanol at high levels becomes toxic, triggering physiological and behavioral changes that could be maladaptive. Egyptian

SCENT

► Neotropical fruit bats like this yellow-shouldered bat have a high tolerance for ethanol in their fruit. In general, though, bats and other mammals avoid fruit that is too ripe because it can be less nutritious and possibly dangerous.

fruit bats fed food containing 1 percent ethanol flew more slowly, which could make them more vulnerable to predators. (Interestingly, Neotropical leaf-nosed bats may not be affected by ethanol in the same way. A study of several leaf-nosed fruit- and nectar-feeding bats found no strong effects of ethanol consumption on either flight performance or echolocation.) Second, high levels of ethanol mean that yeast and other microbes have been hard at work, breaking down and using the sugar and nutrients in the fruit. The more those nutrients are consumed by these microbes, the less is available for the bat, making overripe fruit a poor investment when it comes to energy and nutrient levels.

After seeing the responses of fruit-eating bats to various ethanol levels, researchers made a slight tweak to their experiment. They presented bats with a choice between food containing 0 percent or 1 percent ethanol. Bats that had fasted for 24 hours showed no preference between the two options, whereas well-fed bats continued to avoid the ethanol. In these instances, getting a little tipsy is better than potentially starving, thus influencing the choices bats might make based on scent and taste.

# Sweet and Sulphury Flowers

Bat-pollinated flowers stink. *Strong, fetid, musty, skunky*, and *pungent* are just a few of the words used to describe chiropterophilous (bat-loving) flowers. That's thanks to the production of sulfur-containing compounds, like dimethyl disulfide and dimethyl trisulfide. Sulfur compounds are surprisingly common cues used by vertebrates to find food resources. Albatross and other wide-ranging seabirds appear to cue in on sulphury sea scents (mainly dimethyl sulfide), which helps them orient and hunt in a constantly shifting watery landscape devoid of other useful landmarks. However, sulfur odors in flowers are uncommon and are usually associated with either bats or carrion and house flies.

Many species of pollinating bats love that sulfur smell. Pallas's long-tongued bats and some other nectar-feeding New World leaf-nosed bats have innate preferences for sulfur compounds. Even captive-born bats never exposed to natural floral scents have strong preferences for artificial flowers containing dimethyl sulfide. The presence of sulfur compounds across many different types of bat-pollinated plants is a prime example of plant-animal coevolution, where interacting species influence each other's evolution. Floral scents like sulfides help bats locate certain species' flowers in cluttered forest environments, where they can then pick up and distribute pollen among individual plants. This benefits the plants by helping mix up their genetic material and produce offspring with greater fitness. Bats benefits by getting nutrition in the form of high-energy nectar. Night-opening flowers that emit sulfuric compounds and are white or dull colored, rich in nectar, and tube- or cup-shaped likely rely on bats as their main pollinators.

A fascinating example of this coevolution between bats and plants can be seen with the kapok tree, which is distributed in rainforests of subtropical regions of West Africa and South America. The kapok tree is distinguished by large, sprawling buttress roots, thorny bark, and tall umbrella-shaped crowns. In South America, the trees have clusters of small white flowers, dominated by odor compounds like dimethyl sulfide and dimethyl disulfide. But when scientists analyzed the floral scents of this same species in Africa, they were surprised to find very few sulfur notes in the floral bouquets. In both places, flower-visiting bats appear to be the primary pollinators. Having been separated from each other for millions of years, the two populations evolved different scents, even though

► The shaving brush tree is an example of a bat-pollinated flower.

▲ The flower of the bat-pollinated kapok tree. Kapok flowers in the Paleotropics don't have the same sulphury smell as those in the Neotropics.

◄ Can insect-eating bats like these little brown bats use their sense of smell to find food or roosts? So far, the evidence is unclear.

both populations are bat-pollinated. And while Central and South American nectar-feeding bats are attracted to sulfur smells, that may not be the case for all nectar-feeding bats worldwide. When cave nectar bats inhabiting subtropical Asia were presented with kapok flowers that were artificially scented with dimethyl disulfide, they did not demonstrate a preference for the sulfur smell. Instead, the Asian bats seemed to avoid sulfur-treated flowers.

# Not Just Fruit Bats

You know the old saying "you can't teach an old dog new tricks"? Well, you also can't teach an insect-eating bat to hunt using smell. Early in my PhD program, I wanted to see if the insect-eating Mexican free-tailed bat uses odor cues during foraging and social interactions. I trained wild-caught Mexican free-tailed bats to associate a certain smell with a tasty mealworm reward. Using a similar method to the one I used in my fruit bat experiments in Panama, I recorded how often bats correctly chose to approach the reward. But after hours of training, the bats rarely approached the odor to get their reward and seemed to get worse with training! It wasn't until I talked to other bat researchers that I learned that some bats are difficult to train and trying to get insect-eating bats to care about odor is challenging. This difficulty has led to the assumption that insect-eating bats don't use odors during foraging. Although one study reported that the greater mouse-eared bat uses odor cues to find beetles buried in the sand, other field studies have revealed that acoustic cues are almost always more important than smells for hunting insect-eating bats.

There is still quite a bit to learn about the nuances of how bats use smell for foraging decisions and hunting. But while I think the use of smell in insect-eating bats is underestimated, it seems unlikely they use it as a primary cue for finding food the way that fruit- and nectar-feeding bats do. However, across almost all bats—regardless of if they eat fruit, nectar, insects, or meat—odors are thought to play some role in behaviors like mate selection, kin recognition, and other social interactions.

# VIEW

Chapter 4

Blind as a bat? Not exactly. Despite what you might have heard, all bats can see. Even bats that rely primarily on echolocation for getting around and capturing prey still use vision for at least some things, whether its spotting movement of insects in cluttered areas or using light as a cue to know when it's time to leave the roost at night. In the same way that bats' abilities to use sound and smell are closely linked to their ecology, so too is their use of vision.

▸ Not only are bats *not* blind, but some have senses humans only dream of. This common vampire bat has heat "vision."

# 20/20 Vision?

What does it mean to have "good vision" anyway? Visual stimuli, such as light and colors, are perceived primarily by the retina that lines the back of the eye. Special light-sensitive nerve cells in the retina called photoreceptors react to incoming light and send electrical impulses via the optic nerve to the brain, where these signals are integrated and processed by the visual cortex. Deep within the inner layer of the retina are the retinal ganglion cells, whose axons form the optic nerve. The optic nerve is then responsible for sending this visual information to the brain.

Vertebrates have two main types of photoreceptors: rods and cones. Rods are super sensitive to light in the environment, making them most useful in dim light and dark conditions. Cones need brighter, more intense light levels to be activated, making them useful for seeing in daylight conditions. Different types of cone cells respond to different wavelengths of light, allowing animals to see in color. Visual acuity, or how well the eye can resolve detail of an image at different distances, also depends primarily on the cones, which is why our ability to see details and colors decreases with light levels.

When we stand in a doctor's office reading letters off a Snellen chart of increasingly smaller and smaller letters, what we are measuring is our visual acuity. It's challenging to measure visual acuity in nonhuman animals, since we can't exactly ask them to read letters off a piece of paper. Instead, scientists measure anatomical aspects of an animal's eye that are expected to correlate with acuity. One measurement used is the axial diameter of the eye (or eye length), which determines the size of a retinal image. If overall eye shape is kept constant, any increase in eye length will result in a larger retinal image. Animals with longer eyes are predicted to have better visual acuity than animals with shorter eyes. Visual acuity can also be estimated by analyzing the overall number and density of retinal ganglion cells in the back of the retina, with higher densities correlating with better overall acuity.

Visual acuity is described in two ways: cycles per degree or minimal separable angle. The more cycles per degree or the smaller the minimal separable angle, the better an animal is at resolving detail at a given distance. Using anatomical measurements of the retinal ganglion cells, the rufous horseshoe bat—an insect-eating bat with small eyes known to rely on its sophisticated

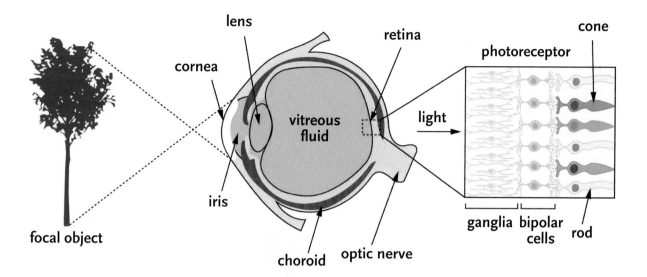

Labels: lens, cornea, retina, cone, photoreceptor, iris, vitreous fluid, light, ganglia, bipolar cells, rod, focal object, choroid, optic nerve

echolocation for hunting—is estimated to have low acuity (about 0.35 cycles per degree), whereas the larger eyed, partly carnivorous Australian ghost bat has an estimated acuity of about 2 cycles per degree. In daylight conditions, humans have about 60 cycles per degree of visual acuity.

Another way to measure a bat's visual acuity is the optomotor response. A type of behavioral test, this is the closest that scientists can get to asking a bat to read letters off a chart. The bat is placed in a device that consists of an inner tube surrounded by a revolving drum. Black and white stripes of varying thicknesses and gradients are painted on the inside of the revolving drum. The width between the black bands correspond to cycles per degree from the bat's point of view. If a bat moves its head to track the stripes as they rotate around the drum, that indicates that the bat can discriminate the pattern.

Optomotor response tests combined with morphological measurements confirm some general patterns about bat sight. Gleaning bats, who hunt by plucking insect prey from surfaces, tend to have larger eyes and better acuity than aerial hunters. At best, aerial insectivorous bats, like the little brown bat and Daubenton's bat, can detect objects of 5–9 cm from a maximum of about 1 m away. While that's not bad, these bats mostly feed on much smaller prey items, meaning they are unlikely to be using vision during hunting. In contrast, the gleaning brown long-eared bat and pallid bat can make out objects 0.9 cm or smaller from a distance of 1 m.

▲ Diagram of the mammalian eye. Light enters the eye through the lens and reacts with the retina in the back of the eye, triggering electrical signals in the optic nerve. Behind two layers of cells are the cone and rod photoreceptors, which detect different levels and wavelengths of light. Cones are named for their conical shape and are responsible for color vision, whereas rods are more oblong and detect the presence/absence of light.

# Seeing (or Not Seeing) Red

Bats can see different shapes and sizes reasonably well, depending on the shape of their eye and their general habits. But what about colors? Rod and cone photoreceptors contain special visual pigments, most notably proteins called opsins. The different types of opsins are named for the wavelength of light to which they are most sensitive. Mammals have two types of cone opsins: long-wavelength-sensitive (LWS) and short-wavelength-sensitive 1 (SWS1). L-cones (with LWS cone opsins) are sensitive to colors like yellow and red, whereas S-cones (SWS1 cone opsins) are sensitive to colors like violet and blue. Most mammals can only distinguish between red and blue colors and can't discriminate in-between colors like green. Humans and Old World primates are exceptions to this rule thanks to a long-ago gene duplication that enables discrimination between medium-wavelength greens and long-wavelength reds.

Bats have both cone and rod photoreceptors, meaning they can discriminate some colors. Genomic analyses of the genes that code for the different opsin proteins provide insight into exactly what colors bats can see. All bats examined so far have functional LWS opsin genes, with most species having highest sensitivity to light wavelengths around 555–560 nm (orange/red). Long wavelengths of light like this are better for maximizing visual information under low-light conditions—perfect for animals like bats that are generally most active after sunset. While some bat species have maintained functional copies of SWS1 opsin genes and can discriminate between red and blue, others have lost function in these genes. Most of these colorblind bats are insect-eating bats with particularly sophisticated echolocation calls, suggesting potential trade-offs between vision and echolocation in bats. However, it's not just insect-eating bats that have lost the ability to see blues and purples; some fruit bats also have nonfunctional SWS1 genes.

Some bat species can see ultraviolet light, very short wavelengths that humans can't see. The nectar-feeding Pallas's long-tongued bat has behaviorally demonstrated an ability to discriminate colors with wavelengths as low as 310 nm (the approximate cutoff between visible and ultraviolet wavelengths is about 400 nm), but it is unable to discriminate between green and orange. Being able to detect ultraviolet wavelengths may be particularly advantageous at dawn and dusk and, in flower-visiting bats, for detecting ultraviolet reflectance of petals.

◄ CLOCKWISE:
Mexican funnel-eared bats have small eyes and rely primarily on echolocation to navigate.

Spectacled flying foxes and other pteropodid bats have large eyes and don't echolocate.

While they do have large eyes and can likely see well in dim light, cave-dwelling hairy-legged vampire bats have mutations in their genes that prevent them from discriminating the color blue.

The flower-visiting long-tongued bat can see some very short wavelengths of light, down to about 310 nm (ultraviolet light).

VIEW

What drives variation in color vision among bat species? One main hypothesis is that it is linked to the type of echolocation bats use or if they echolocate at all. While the evolution of the most advanced types of echolocation calls does usually coincide with loss of short-wavelength color vision, some non-echolocating fruit bats also have a nonfunctioning SWS1 gene. Another hypothesis centers around choice of roost, based on the observation that cave-roosting bats (both echolocating and non-echolocating) seem more likely to have lost the SWS1 gene and the ability to see purple. Short-wavelength vision is not as useful once light levels fall below a certain threshold, so for bats that spend most of their life in the dark of the night or a cave, there was no survival benefit to maintain those genes. Based on current studies, this hypothesis seems well supported, although there are still many species of bats whose opsin genes have yet to be sequenced.

Another surprise from the bat world? At least one bat species may be vying for a spot next to primates as the only mammals known to discriminate three types of colors. Fischer's pygmy fruit bat was found to have a duplication of the LWS opsin gene, similar to the evolutionary process that resulted in expanded color vision in some primates and humans. At this point it remains unclear exactly if and how that duplication might affect this bat's vision.

## Follow the Light

Both eye morphology and behavior provide strong evidence for functional—and in some cases very good—vision in bats. However, even the most sharp-eyed bat would have difficulty seeing on a dark, moonless night or tracking a small, rapidly moving insect with its eyes. How important is vision in the daily lives of bats, and how do bats integrate visual cues with echolocation?

Light cues play important roles in regulating bat circadian rhythms and activity patterns. Circadian rhythms are what sets the body's internal clock, helping regulate things like hunger, hormone levels, and sleep. Daylight helps bats time their nightly emergence from roosts, and light is an important visual cue for orienting bats to cave exits. When insect-eating bats are placed in a special Y-maze where the bat must crawl or fly through one of two tunnels to escape, they show strong preferences for lighted tunnels even when the end of

the lighted tunnel is blocked with clear glass that is detectable by echolocation. This preference for a lighted exit also varies across the day, with bats showing stronger preferences for lighted tunnels corresponding to the time when they would normally be leaving the roost.

Bats combine all sorts of sensory information from their environment to navigate and make decisions. In some scenarios, one sensory cue may be more useful or convenient than another. In addition to using vision to guide their escape response, bats will also take advantage of visual cues when there's enough light to do so. Even when temporarily deafened using earplugs, short-tailed fruit bats and spear-nosed bats were both able to successfully navigate a semi-lit obstacle course consisting of long, white strips of cloth. There are also scenarios where visual information is more useful than acoustic information. In cluttered forest environments, sounds bounce around more than in open air,

▲ While short-tailed fruit bats can and do use echolocation, they also use vision to navigate around large obstacles. This is especially important when acoustic cues might not be available or reliable.

making it more difficult for bats to isolate returning echoes. When presented with different moths in identical conditions, northern bats were more likely to attack white moths than black moths, suggesting that visual contrasts were helping guide bats during hunting (although they still used echolocation calls during these attacks). Echolocation calls can also help provide complimentary information to bats that are frequently active during the day. Egyptian fruit bats regularly use their echolocation clicks when foraging, landing, and flying near obstacles like fruit trees, even in full daylight conditions.

# Seeing the Way Home

There are some cases where echolocation cues are unlikely to be helpful, like over long distances. Sound dissipates in air much faster than light, which explains why we can see city lights from far away but can't hear city traffic at the same distance. Visual cues are important for long-distance homing and navigation in bats, and studies have shown that bats use both large landscape features like mountains and even celestial cues like stars to guide their movement.

Homing is an animal's ability to return to home or a known location such as a roost or specific foraging or hibernation site after being displaced. In the

▲ Big brown bats show evidence of being able to use both sunset glow and the Earth's magnetic field to help them orient over long distances.

1950s and 1960s, researchers would capture and mark bats at one location (such as a roost) and then release them somewhere unfamiliar. Then they counted how many of the marked bats successfully returned home and how long it took the bats to find their way back. Blindfolded bats, when released from an unfamiliar location far outside their home range, were less likely to make it back to the roost than bats that were not blindfolded. This suggested some reliance on vision to orient over long distances, but these findings were difficult to interpret, as there are many reasons why bats might not make it home, the exact paths taken by bats were unknown, and bats were often pretty good at removing the blindfolds on their own.

More recent evidence for use of visual landmarks comes from studies of GPS-tagged bats and technology that lets researchers reconstruct the exact flight paths bats use to get from one spot to another. In Israel, researchers tagged Egyptian fruit bats with GPS units and then released them from a crater that was about 50 miles from their cave roost. One group of bats was released from the top edge of the crater, whereas another group was released from the bottom of the crater. Both groups of bats successfully found their way back to the roost, but bats that were released from the bottom of the crater took longer and less direct paths back than the bats released from the top of the crater—suggesting that bats rely on visual information to orient. Bats released at the top of the crater would have had immediate access to landscape cues such as mountains or the sea to orient, but bats at the bottom of the crater lacked these useful landmarks.

Sun position and post-sunset glow are also useful visual cues to help bats orient on the landscape. When individuals from a colony of big brown bats were released at an unfamiliar location, they all took off in the same direction when released individually at sunset, but they flew in random directions if released after dark. Migratory soprano pipistrelles can use the angle of the sun at dusk to calibrate their internal compass during migration. When bats were experimentally exposed to a 180° rotated angle of the setting sun (by using a mirrored image of the sunset), they shifted their take-off heading by the same amount as compared to control bats not exposed to this mirror image. In these cases, vision might not provide specific information about objects in the landscape but instead act as general markers for long-distance movements, like ocean navigators using the position of the stars to navigate the seas at night.

# Magnetic Sense

The leaves crunch underfoot as I crest a small hill in the late autumn forest, my dad walking at my side. Growing up, my dad frequently took my sister and me on hikes through our local forests, armed with field guides on native wildflowers and a Thermos full of hot chocolate. On this particular day, he was teaching me an important life skill—how to read a map (and secondarily, how not to get lost in the woods). Armed with a topographic map of the forest and his trusty magnetic compass around my neck, my dad directed me to find our lunch spot, which he had set up and marked about an hour earlier. Flipping open the compass, I rotated around, watching as the red arrow indicating north spun around with my movement. Compasses like this work because the small, magnetized needle floating at the center lines up with the Earth's magnetic field, indicating the direction of magnetic north and south.

As a human, I have to rely on this little device to point me in the right direction, but if I were a bird, I might not need its help. That's because some animals (including sharks, sea turtles, and salmon, as well as birds) can detect the Earth's magnetic field without help. There's good evidence that bats can too.

So far, evidence for magnetic orientation is limited to two species: the big brown bat in North America and the Chinese noctule in Asia. In one study, researchers exposed two groups of big brown bats to a rotated magnetic field, using a device called a Helmholtz coil that creates a local magnetic field that can be manipulated in different directions. If bats are using the Earth's magnetic field like a compass, then they should be able to sense the direction of north and south and orient themselves accordingly. After hanging out in the Helmholtz coil for about 90 minutes, bats were then released 20 km north of their roost site and tracked using radio telemetry. Bats that were exposed to a magnetic field that had been rotated 90° oriented toward the east, whereas bats that had been exposed to a 270° rotated magnetic field turned west, in line with the predicted change of direction. Meanwhile, control bats that were not exposed to a manipulated magnetic field went in the correct direction of their roost.

Similar results were observed in a laboratory-based behavioral experiment with Chinese noctules. Wild-caught bats hanging comfortably in a basket were exposed to a magnetic field with a reversed polarity, meaning that if we stuck our magnetic compass in the reversed field, the arrow would point south. Prior to their exposure to the altered magnetic field, the bats preferred roosting at the northern end of the basket. After the horizontal poles were reversed, they switched, instead showing a preference for the southern end of the basket.

Exactly how animals like birds and bats can detect the magnetic field is still a bit of a mystery. One possible mechanism is that these animals have specialized photopigments that react to light in a way that is sensitive to the magnetic field. Another possibility involves magnetite, a form of iron oxide with magnetic properties that cause it to align with a magnetic field. Free-moving magnetite in a cell might act like miniature arrows, lining up with the direction of the magnetic field the same way they do in a compass. Magnetite has been identified in bacteria and more recently in larger animals, such as birds and fish. When big brown bats were tested in a pulse remagnetization experiment designed to reverse the polarity of these tiny magnetic particles and released away from their home roost, they changed their orientation relative to the roost. However, researchers in this study were not able to isolate exactly where these particles might be in the bats. Another study looked at the brain tissue of six bat species and found evidence for soft magnetic particles, lending further support to the idea that bats might use magnetite to detect the Earth's magnetic field.

## Turn Up the Heat

In addition to sensing ultraviolet light and the Earth's magnetic field, let's add another superpower to the list of bat senses: heat vision. While using sound and smell to locate prey, vampire bats can also detect thermal radiation via infrared-sensitive receptors on their nose. This so-called heat vision is not image-based—not in the way that a thermal camera uses thermal radiation to create pictures. Instead, these infrared-sensitive receptors are thought to help bats locate blood vessels under the skin of their prey, which bats can sense from up to 20 cm away. Infrared and thermal sensing in vertebrates is relatively rare, thus far found only in three groups of snakes (boas, pythons, and pit vipers) and vampire bats—the only mammals known to have this unique sensory ability.

Within the mammalian face is a large cranial nerve called the trigeminal nerve. This nerve has branches extending to the eye, cheek, and jaw and is

responsible for sensing touch, pain, and temperature in the face. All three species of vampire bats have unique features associated with this nerve that are not found in other mammals, particularly at the molecular level. In mammals, burning sensations (like when we eat something spicy) are due to the activation of a special receptor molecule called TRPV1; this same molecule also helps signal our brain when we touch something too hot. In humans and many other animals, that temperature threshold is around 43°C (110°F). Vampire bats have both the normal version and a second, slightly modified version of TRPV1 that is shorter and has a heat threshold much lower than the original, activating when sensing heat around 30°C (86°F). The normal form of TRPV1 is produced in the spinal cord, whereas the modified form is only produced in the vampire bat's facial nerves. So far, this specific TRPV1 variation has only been confirmed in common vampire bats, though it is likely also related to infrared sensing in the other two species as well.

▲ Common vampire bats can sense heat using special thermoreceptors on their nose.

# BITE

## Chapter 5

Among mammals, it could be argued that bats as a group have the largest range of diets. While the majority of the world's bats feed mostly on insects, there are bats that specialize on eating fruit, nectar, pollen, spiders, venomous scorpions, fish, birds, rodents, blood, and even other bats. Some, like pallid bats, are omnivorous jacks-of-all-trades, eating a wide variety of foods, whereas others, like vampire bats, have become so specialized they literally can't survive on anything but one food item.

▶ A Mexican long-tongued bat flies in to feed from an agave flower.

# Nectar-Feeding Bats

A variety of bat species in both the New World and Old World tropics feed on the nectar and pollen of flowers, while also serving as important pollinators for many plant species. Getting to the tasty nectar at the bottom of a flower can be a lot of work, and nectar-feeding bats have evolved elongated noses and tongues that make it easier to reach into flowers. Fruit-eating flying foxes will also take advantage of flower resources, whereas another group of tiny bats are more specialized on flowers. Blossom bats and their relatives (dawn bats and long-tongued fruit bats) have long tongues and noses that help them extract nectar from flowers. Unlike New World nectar-feeding bats, which hover above the flower during feeding, blossom bats land on branches containing flowers and use their nose to help push the flowers open. Nectar-feeding bats in the tropics of Asia, Africa, and Australia tend to be much less specialized on one type of flower or fruit.

## Mops and pumps

The nectar-feeding bats in the tropics of Central and South America have some highly specialized morphology for feeding from flowers. In many ways, these small nectar-feeding bats operate very similar to hummingbirds; they can hover as they insert their head into a flower opening, all while extending an elongated tongue. As in Pallas's long-tongued bat, at the tip of the tongue is a collection of projections called papillae. When at rest (inside the bat's mouth), these triangular papillae form a scaled pattern. When the bat sticks its tongue out to the maximum length, blood rushes into the papillae and causes them to become erect and spiky. This maneuver turns the bat's tongue into a nectar mop, with nectar getting trapped between these spiky projections. Bats with this type of tongue feed by rapidly lapping their tongue in and out of their mouth during feeding, a behavior also sometimes called viscous dipping.

Not all nectar feeders have these tongue spikes. A group of related nectar-feeding bats in the genus *Lonchophylla* have a slightly different strategy for slurping nectar. Instead of elongated papillae, they have deep grooves that run along the sides of the entire length of the tongue. When feeding, *Lonchophylla* bats keep their tongues just below the nectar surface, and the nectar is pumped upward into the bats' mouths via a combination of cell pumping movements and capillary action.

Why did these groups of closely related bats evolve such different ways of feeding on the same thing? Although initial observations of this tongue-pumping method suggest it might be more efficient at pulling nectar from flowers than spiked-tongued lapping, it might be related to variation in flower nectar. Thick, viscous nectar might be more easily harvested using brush-tipped tongues, whereas tongue grooves are better for flowers with more dilute nectars.

## How long can you go?

Brush tips and grooves aren't the only fascinating thing about nectar-feeding bat tongues. In one study, researchers measured the operational tongue length of a bunch of nectar-feeding bat species. They trained the bats to feed from glass tubes of different sizes to see just how far they can stick their tongues to drain the sugary treat. Turns out, it's far. Of the bats in the study, the aptly named Mexican long-tongued bat could stick its tongue out about 7 cm, longer than its entire body length. In comparison, the fruit-eating short-tailed bat could only extend its tongue 2.4 cm.

But hold your applause! In 2005, a new species of nectar-feeding bat was identified in the cloud forests of the Ecuadorian Andes. The tube-lipped nectar bat shares the forest with two similar nectar-feeding bat species: the tailed tailless bat (and, yes, that's its common name) and Geoffroy's tailless bat. When Nathan Muchhala, one of the researchers that described the new species, measured the tube-lipped nectar bat's tongue, he found it could extend almost 8.5 cm—more than twice the length of the other two species' tongues and about 150 percent of its own total body length. This bat's tongue is so long that

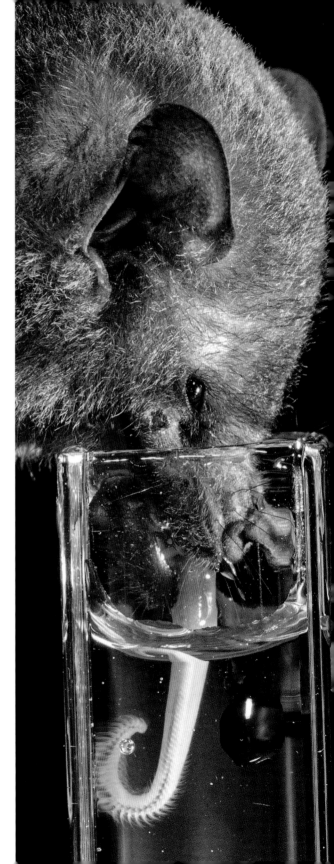

▲ A black flying fox feeds from a fruiting tree.

instead of being anchored at the base of the throat, it passes through the neck into the thoracic cavity, where it's protected by a special sleeve of tissue called a glossal tube, coming to an end between the heart and the sternum. The newly discovered bat is specialized to drink the nectar from just one flower in the cloud forests, *Centropogon nigricans*, with tubular flowers that are about 8–9 cm long, right in the range of the bat's tongue. Diet and foraging behavior analysis suggests that tube-lipped nectar bat is the only pollinator of this plant.

# Fruit-Eating Bats

Fruit is a popular diet option for bats across the world. Nearly all the bats in the Old World family Pteropodidae (flying foxes) rely on fruit as part of their diet, whereas primarily frugivorous bat species in the Neotropics are members of the Phyllostomidae. Fruit preferences among bats depends on a variety of things, including location and seasonality of available fruit; the size of the bat; and its ability to bite, handle, and manipulate the fruit. Flying foxes are generalist fruit eaters, consuming a variety of fruits across their range, including figs, mangos, lychee, tamarind, and guava. Neotropical fruit bats tend to be more specialized in their fruit preferences than Old World fruit bats.

# Short-faced bats

What a bat eats is strongly influenced by how it eats—specifically, how its skull, teeth, and face muscles let it chomp, crunch, or suck at a food like fruit. Imagine the difference between biting into a crisp, crunchy apple and a juicy, ripe peach. In general, bats with bigger skulls can bite down harder than bats with small skulls. The stronger the bite force, the harder the fruit a bat can eat. Figs, a popular fruit for bats across the world, are generally considered a hard fruit, whereas bananas and papayas are soft fruits.

Despite being smaller than flying foxes, which frequently eat hard fruits, many species of New World fruit bats have adapted to also feed on hard fruits like figs. Species that eat hard fruits have evolved short, stocky skulls that are more efficient at producing the high bite force needed to hold, process, and chew these fruits. When paired with a strong temporalis muscle—the muscle that fans across the temple and attaches to the lower jaw—the square, stocky skulls of bats like the Jamaican fruit-eating bat can deliver the forceful bite needed to pierce through the tough outer skin of a ripe fig.

The wrinkle-faced bat is an extreme case of skull shortening in bats. Perhaps the most enigmatic and bizarre bat I have ever encountered, the wrinkle-faced bat's face is covered in naked flaps and folds of skin that create its namesake appearance. These wrinkles cover a squat face with a very short nose, like that of a French bulldog or pug. Wrinkle-faced bats feed on a variety of different fruits across a range of hardness. Observations revealed that they use the same deep, sideways bite when feeding on hard or soft fruits. Wrinkle-faced bats also have very high bite force relative to their body size, producing a bite 20 percent stronger than that of bats of similar size. By having such a strong bite, wrinkle-faced bats can take advantage of an array of fruit depending on availability. This can be handy during times of drought or other environmental stress when softer fruits might not be available, giving bats that can feed on hard fruits an advantage.

▲ A Jamaican fruit-eating bat bites into a fig.

# Spat!

Fruits vary in their nutrient compositions, particularly when it comes to fiber. Although high-fiber fruits tend to be lower quality in terms of nutrients like protein and fat, they are often abundant and easy to find in the environment, making them a preferred food of many fruit-eating bats.

Figs are a great example of a high-fiber fruit. In addition to having strong bites, bats that feed on a lot of figs have other behavioral adaptations that help them deal with all that fiber. After taking a bite, fig-eaters spend more time chewing their food than bats that consumer fruits with less fiber. During all this chewing, the bats use their tongue and cheeks to collect the pieces of seeds, fruit skin, and pulp into a squishy, sticky ball. After squeezing all the juice out of this ball, called a spat, they spit it out. This behavior is seen in both Old World and New World fig-eating bats. This extra chewing time helps the bats extract as many nutrients as possible without having to deal with the costs of trying to digest all that fiber.

# Beyond fruit and nectar

While some bats that feed on fruit or nectar specialize almost exclusively on that resource, many bats are more flexible about what parts of a plant they will eat. For example, many Neotropical fruit-eating bats will also feed on flowers when available.

By spitting out the seeds when they eat, fruit-eating bats play important roles as seed dispersers in tropical forest environments. However, there is also evidence that some bats feed on seeds. At least two species of big-eyed bats (genus *Chiroderma*) are known to eat fig seeds, ingesting the nutrient-rich inner contents while spitting out the hard outer seed coat.

Eating leaves is one way that bats can supplement their diets, as many leaves are rich in a range of dietary minerals, like calcium, that are important for growth and survival. In some species, leaves may form a large part of the diet, such as in the white-lined broad-nosed bat. Behavioral observations of a small colony of this species found that they regularly consumed leaves throughout the year. Leaf consumption was higher than fruit consumption during the dry season, possibly because of food availability.

Plant foods like fruit and nectar are great sources of sugar and carbohydrates needed for energy, but they tend to be lacking in other important nutrients like protein. As more research is done on bat diets, we are learning that bat food habits might be more flexible than originally thought. For example, Jamaican fruit-eating bats may actively seek out insect prey during certain times of the year, such as during pregnancy, to increase their protein intake. Advances in

▲ In addition to eating fruit, hairy big-eyed bats will also digest fruit seeds.

DNA analysis also revealed that the presumed nectar specialist Pallas's long-tongued bats (one of the bats with the spiky tongues) also consistently feed on insect prey, suggesting they are more omnivorous than previously thought.

# Insect-Eating Bats

The first bats were probably insect eaters, with all other diet adaptations evolving later. Most bats across the world still eat insects, with an estimated 70–80 percent of all bat species consuming insects as all or part of their diet. Insect-eating bats have evolved astonishing hunting behaviors and various specializations to avoid competition with each other and other animals.

The zigzagging acrobatics of bats as they zip through the sky is generally associated with a style of hunting called aerial hawking. During these maneuvers bats catch insects straight out of the air. While some bats can directly grab at insects with their teeth, many species will use their wings and tail membranes to help scoop the insect prey toward their mouth, serving as a handy butt spoon.

Gleaning is the other common strategy for hunting arthropods. Instead of capturing insects out of the air, gleaning bats pluck insects (or other prey) off surfaces such as leaves, tree branches, cliff faces, and the ground. Some species, like Natterer's bat, can pluck spiders from their webs and have special stiff hairs along the edge of their tails that help them sense the spider's location. The tiny disc-winged bat also frequently feeds on jumping spiders, which it captures from the surfaces of leaves. Other bats eat insect larvae, plucking them from leaves and tree trunks. Common big-eared bats in Panama can consume as much as 80 percent of their body mass in arthropods in a single night, using their sophisticated echolocation to detect and eat motionless, sleeping caterpillars.

# Immune to venom

In most animals, the potential for getting stung or injected with venom from something like a scorpion is a strong deterrent. Not so much for the pallid bat and desert long-eared bat. Found in desert habitats on separate continents, both species commonly consume potentially deadly scorpions. Observations of both

▲ Aerial insectivores like this Mexican funnel-eared bat capture insects straight from the air.

◀ Much larger than the funnel-eared bat, golden bats capture large insects like katydids and beetles from leaf surfaces. They can also grab vertebrate prey like small lizards.

▼ These pallid bats in a night roost in southwestern Texas are gleaners that regularly consume venomous scorpions. The bats appear to be immune to scorpion stings.

species find they are frequently stung while attacking scorpions, including in the face. Despite this, the bats never show any evidence of pain, and desert long-eared bats will readily eat the whole scorpion—stinger included.

Pallid bats in western North American regularly consume the Arizona bark scorpion, one of the most venomous scorpions in the United States. The venom of the bark scorpion works by targeting sodium channels in the cells, resulting in pain and muscle paralysis. Compared to mice, pallid bats have differences in their DNA sequences that make it harder for the venom molecules to bind to their sodium channels, making them venom-resistant. At this point it is unknown if the desert long-eared bat has a similar adaptation for dealing with their potentially deadly prey.

◄ Some bats, like this Virginia big-eared bat, use soft, low-frequency echolocation calls to avoid being heard by their prey.

▲ Others, like the naked-backed bat, are very loud.

# The great evolutionary arms race in the skies

The ability to use sound to navigate and detect insects in the dark of the night certainly gives bats the advantage as nocturnal predators. If the prey insect can hear a bat coming, however, it gives the insect a chance to try to escape or, in some cases, interfere.

In insects, the ears consist of a thin external membrane, called the tympanum, that covers internal air sacs. Sensory cells attach to this tympanum and respond to sound frequencies. While many insects cannot hear, those with these structures can. At least four insect orders commonly eaten by bats are known to have a tympanum, including Lepidoptera (moths and butterflies), Orthoptera (crickets), Dictyoptera (mantids, cockroaches, and termites), and Neuroptera (lacewings).

► The tiger moth *Bertholdia trigona* produces high-frequency clicks to disrupt the echolocation calls of hunting bats.

► The tiger moth *Bertholdia trigona* produces high-frequency clicks to disrupt the echolocation calls of hunting bats.

The simplest way to try to avoid a bat attack? Listen for the sounds of a shouting bat. Some of these hearing insects respond to high-pitched, batlike sounds with escape tactics like jumping away, zigzagging flight, and diving or falling away.

Some lepidopterans have taken things a step further and produce their own sounds as a response to bat echolocation calls. Some butterflies in the family Nymphalidae produce clicking noises, which can startle bats during attack. Bats quickly get used to this response, however, and the strategy can backfire, instead acting as an attractive dinner bell for experienced bats. One species of tiger moth, *Bertholdia trigona*, can disrupt the echolocation calls of bats with high-frequency clicking, up to an amazing 4500 times a second. When exposed to these fast-clicking moths, bats don't learn to avoid them—they physically cannot catch them. It's thought the rapid clicking creates a blurred acoustic image for the bat, which can sense that the moth is nearby, but not exactly where it is.

Other tiger moths retain distasteful and even poisonous chemicals from their plant hosts, the same way monarch butterflies acquire toxins from milkweed. In addition to having bright colors that might serve as a warning to avian predators, these moths produce clicks that help bat predators learn to avoid them. In behavior experiments, bats quickly learned to associate these moth clicks with bad-tasting prey, avoiding more than 90 percent of the toxic moths released.

What about insects that can't hear bat echolocation calls? Turns out, they might not be exactly defenseless either. The long hindwing tails of luna moths can act like an acoustic diversion that prevents a bat from grabbing the moth's body. Instead, the bat aims its attack at the edge of the wings. Compared to luna moths that had their tails removed, intact moths showed a 47 percent survival advantage when faced with a hungry big brown bat. The wings of other moth species can absorb some of the ultrasonic frequencies in bat calls, decreasing a bat's ability to accurately detect the moth's location by acting as a type of acoustic camouflage.

# Move Over, Dracula

Although they seem like they are only born from spooky legends, vampires are very much real. Of the hundreds of bat species across the world, only three have taken up the blood-feeding lifestyle. And when it comes to adaptations for feeding on a specialized liquid diet, no other bats have taken things as far as the vampires. Sanguivory, or feeding on blood, has led to many unique adaptations in tooth shape, saliva production, limb morphology, digestion, and kidney function.

Vampire bats only inhabit the tropical and subtropical regions of Central and South America. Hairy-legged and white-winged vampire bats feed mostly on avian blood, whereas the common vampire bat is a mammalian specialist. Prior to widespread human settlement, vampire bats likely fed on a variety of wild animals, including tapir, deer, doves, turkeys, parrots, and even other bats. Following years of human settlement and agricultural expansion, however, vampire bats have readily adjusted their prey preferences for domestic animals like cows, sheep, pigs, horses, chickens, and (only occasionally) humans. Of the three species, hairy-legged vampire bats seem to be the pickiest eaters, strongly preferring avian blood to that of mammals. However, a recent study using genetic tools to detect prey types in vampire bat poop found some evidence that hairy-legged vampire bats might occasionally feed on mammals.

While we tend to picture vampire fangs as replacing the canine teeth in humans, vampire bats don't just have sharp canines. In fact, the main vampire bat fangs are sharpened incisors, the front two teeth, which the bats use to create very small (about 3 mm wide) diamond-shaped cuts in their prey. Because

▲ The common vampire bat feeds exclusively on mammalian blood.

◀ The hairy-legged vampire bat primarily consumes the blood of birds.

feeding on blood doesn't require any chewing, the number and size of the rest of the vampire bats' teeth are reduced compared to those of other bats. Remarkably, their teeth seem to stay razor sharp even as vampire bats age, with dulled teeth only observed in some of the oldest known individuals (17 years in the wild).

One thing *Dracula* gets wrong? Vampire bats don't suck blood. Instead, they use their specialized tongue to lap at blood as it trickles from the bite wound. Somewhat like the tongues in nectar-feeding bats, common and white-winged vampire bats have grooves that stretch along the length of the tongue, whereas hairy-legged vampire bats lack this tongue groove. However, all three vampire bat species have a deep groove in their lower lip for sticking out their tongue during feeding.

Depending on who you ask, vampire bats could also be considered venomous mammals. Venom is defined as a toxin (or combination of toxins) delivered to a wound via a sting or bite. Toxins don't necessarily have to cause pain or death, but rather produce some kind of physiological change or response in the envenomated animal. Containing a cocktail of different proteins and molecules that affect blood flow, vampire bat venom helps make it easier for them to feed from the small wounds they make on their prey.

The most well-known venom is draculin (named for Count Dracula), a small protein that works as an anticoagulant, preventing blood in the wound from clotting. In addition to this and similar proteins that prevent clotting, vampire bat saliva also contains special molecules called *Desmodus* salivary plasminogen activators. When vertebrates are wounded, blood forms a clot to stop excessive blood loss and promote healing. Eventually, that clot needs to be broken down. Plasminogen activators are enzymes that prompt the formation of molecules called plasminogens, which go in and chop up the blood clot. With their specialized activators, vampire bats have commandeered this process, using the victim's own plasminogens to prevent clotting while they are feeding. Other molecules in their saliva provide additional important functions to keep vampire bats healthy and happy, including molecules that kill microbes and others that make blood vessels widen, increasing the flow of blood to the wound.

## Water and protein overload

Although the volume of blood a vampire bat ingests during feeding is small—only about 15–20 ml or 1 tablespoon—it can make up nearly half of the bat's body mass. To accommodate this relatively large amount of fluid, vampire bats have several stomach, intestine, and kidney adaptations. Vampire bat stomachs have a large tubular extension, called a cecum. Ingested blood goes straight to the small intestine, and the cecum acts a like an overflow drain, housing the extra blood. This configuration both helps allow for rapid ingestion of blood and acts as a blood storage unit for later regurgitation and sharing with fellow bats.

Blood is not exactly the most nutrient-rich food source available, made of about 90 percent water. Vampire bats have adapted to this high volume by getting rid of water as soon as possible. They start peeing out excess water almost immediately after they begin feeding, even while still licking from their victim's wounds. Roughly 25 percent of the ingested blood weight is passed in the first few hours following the start of feeding. Aside from the water, the rest of the blood meal is about 94 percent protein and only about 1 percent carbohydrates. Blood also contains large amounts of iron, and vampire bats can consume nearly 800 times the amount of iron consumed by humans without ill effects. Overdosing on iron can lead to poisoning, largely affecting the intestines and liver. Researchers have found that vampire bats have lost one of the genes associated with iron uptake in the intestines. Because of this missing gene, vampire bats don't take up iron as efficiently as other mammals, making it easier for excess iron to be safely excreted from the body.

Loss of another gene helps vampire bats deal with their high-protein diet. An enzyme called trypsin is responsible for the breakdown of dietary proteins and helping to absorb that protein into the body. To prevent trypsin from going too wild and breaking down their own cells, most mammals also produce trypsin inhibitors that slow down the enzyme's work. Vampire bats have lost a gene that produces a trysin inhibitor, leading to higher trypsin activity and better breakdown of protein in vampire bat guts.

# Carnivores: We Want the Meat!

Mammals in the order Carnivora, which includes animals like dogs and cats, consume vertebrate prey as a primary food source—or at least used to, in the case of domesticated dogs. Eating mostly meat is actually pretty rare across the mammalian tree of life, regularly occurring in only three of the twenty-six orders of mammals: Carnivora (canines, felines, bears, raccoons, seals), Cetacea (dolphins, porpoises, whales), and Chiroptera (bats). Carnivorous bats are found in a few families, including Phyllostomidae (New World leaf-nosed bats), Megadermatidae (false vampire bats), and Nycteridae (hollow-faced bats).

Meat-eating bats are primarily perch-hunters, hanging and listening for the sound of scurrying prey before swooping down to hunt. The bats in these families are distributed across tropical and subtropical regions of the world. For a long time, scientists assumed there were no carnivorous bats in temperate regions. Bats have since proven us wrong, however, with at least three species of bats in the family Vespertilionidae now having been observed hunting birds in Europe and Asia.

Diets across carnivorous bats span a whole spectrum of prey, including rodents, lizards, frogs, songbirds, parrots, hummingbirds, and even other bats. But among the meat-eating bats, none are exclusively carnivorous, as most will opportunistically or seasonally consume other types of foods. Usually this means insects or other arthropods, although fruit and nectar have also been recorded in the diets of carnivorous bats. Many false vampire bats (Megadermatidae) eat rodents throughout the year, but increase the number of insects eaten during the wet season. The rare ghost bat of Australia has a pretty diverse carnivorous diet, feeding on a mix of rodents, reptiles, and birds during the dry months. Ghost bats are also reported to eat at least ten species of insectivorous bats and will even eat the highly invasive and potentially toxic cane toad.

These diverse diets might explain why there is no singular morphological or behavioral feature that fully separates carnivorous bats from other bats. In general, carnivorous bats tend to be larger than related bats and have lower wing loading and aspect ratios that allow for maneuverable flight, reduced molars, and bigger brains.

▲ Woolly false-vampire bats are generalist carnivores known to consume rodents, lizards, birds, other bats, insects, and even fruit.

## Frog hunters

Among the most well-studied carnivorous bats is the fringe-lipped bat. These unique Neotropical bats have distinctly lumpy projections on their chins, giving them their name. In Panama and some other parts of their range, fringe-lipped bats have become particularly adept predators of frogs. Although they can and do use echolocation, the bats let their food call to them. Fringe-lipped

► The fringe-lipped bat is named for its lumpy lower lip and chin.

bats listen for the calls of male frogs displaying to attract females in ponds, using the combination of croaking sound and ripples caused by the expansion of the frogs' vocal sacs to make their final attacks. These bats can also recognize and discriminate the calls of different frog species and quickly learn to avoid the toxic or unpalatable frogs, with an exceptional memory for sounds.

To study the hunting behaviors of these bats in the lab, researchers at the Smithsonian Tropical Research Institute used positive reinforcement training on wild-caught bats, rewarding them when they approached during certain sounds (old cell phone ringtones). Wild bats that had been trained on these ringtones, released, and then recaptured years later (up to 4 years after initial training) were found to remember the artificial tones, swooping down to get their little treat after hearing the right cell phone chime. Being able to remember a certain sound, even years later, probably helps these bats take advantage of rare prey.

While much of the interest in the hunting behaviors of fringe-lipped bats has focused on their adaptations to hunt frogs, some populations are thought to be only seasonally carnivorous, relying on insects and arthropods at different times throughout the year.

# Birds on the brain

Among the bird-eating bats, most hunt by sneaking up on sleeping birds. The Neotropical spectral bat hunts birds year-round, munching on prey ranging in size from about 20 g all the way to 150 g; the bat itself weighs between 130 and 200 g. These large bats seem to prefer birds such as doves, cuckoos, trogons, ani, and motmot. Many of their preferred prey either roost in groups or have particularly strong odors, and it's thought that spectral bats might use odors to track down prey.

Scientists generally assumed that the size difference, wing shape, and wing loading of bird-eating bats limited them to gleaning attacks on stationary or sleeping birds. In the early 2000s, however, researchers in Spain published the first reports of a vespertilionid bat, the greater noctule bat, feeding on migrating birds. The initial evidence was considered slightly controversial, as the conclusions relied primarily on the presence of feathers in bat poop. But later studies confirmed these findings, providing evidence of bird bones in fecal samples. What is particularly remarkable about these findings is that the greater noctule bat is not very well designed for gleaning flight, having a morphology better adapted for capturing and consuming prey on the wing.

A later study took advantage of stable isotope analysis to understand if and when these bats are feeding on birds. Stable isotopes are alternative forms of elements, such as carbon and nitrogen, that don't decay very quickly. When an animal eats something, whether it's a plant or another animal, those stable elements get incorporated into their body tissue, such as fur, skin, and teeth. By specifically looking at the ratios of nitrogen isotopes in animal tissue,

 In addition to long feet and claws, greater bulldog bats also have long tail membranes and calcars (heel bones), which they use to help them handle their slippery prey.

scientists can estimate where in the food web that species sits. By analyzing isotope levels in the blood of greater noctule bats during different seasons, researchers confirmed that birds make up a large part of their diet, particularly during the fall. It appears that greater noctule bats take advantage of the large number of songbirds that migrate across Europe each year, possibly capturing them right out of the air.

## Go fish!

If I ask you to think of a marine mammal, chances are the first things to pop to mind are whales, dolphins, and maybe seals. Good news! We can add a bat to that list, the aptly named fish-eating myotis. These medium-sized bats (25 g) are found only in coastal areas of Baja California and the Gulf of California, where they take nightly hunting trips over the ocean looking for small floating crustaceans and fish. Marine prey is the main source of food for these bats, though they will occasionally supplement their diet with terrestrial insects.

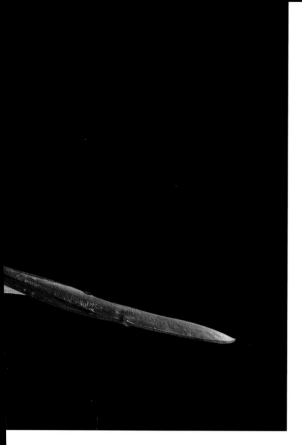

Fishing behavior has evolved several times in bats, and examples of fishers are found across the globe. One thing these bats have in common is large and elongated hind feet with sharp claws that they can rake through the water to catch fish. The greater bulldog bat is the largest of the known fish-eating bats, with sharp claws that can stab through the body of a fish while hunting. Fishing bats also have high concentrations of tactile skin cells on their feet and tail membranes, which help them detect when they've clawed a fish just under the water.

Not surprisingly, bat echolocation doesn't work very well underwater and locating fish is challenging. When fish come up toward the surface, ripples form along the top of the water. Greater bulldog bats can recognize these disturbances as potential prey and come in to rake their long claws through the water. Smaller fish-eating bats, like the European long-fingered myotis, are not quite as sophisticated in their prey discrimination, as evidenced by common attempts to capture inedible objects like plants off the water surface.

▲ The impressively long feet and claws of the greater bulldog bat are used to catch fish from just below the water's surface.

# FLIGHT

## Chapter 6

Those of us who are theater fans or former theater kids are familiar with jazz hands—palms out, with the fingers spread as wide as they go, maybe even with some bonus shimmies if we're feeling fancy. Now imagine those fingers growing and growing, until they extend more than the length of your body. Add skin membrane stretched between the tips of each finger and stretching from the tip of your pinky finger to your ankles. Voila! Now you are a bat!

▶ Van Gelder's bat is an insect-eater that lives in Belize.

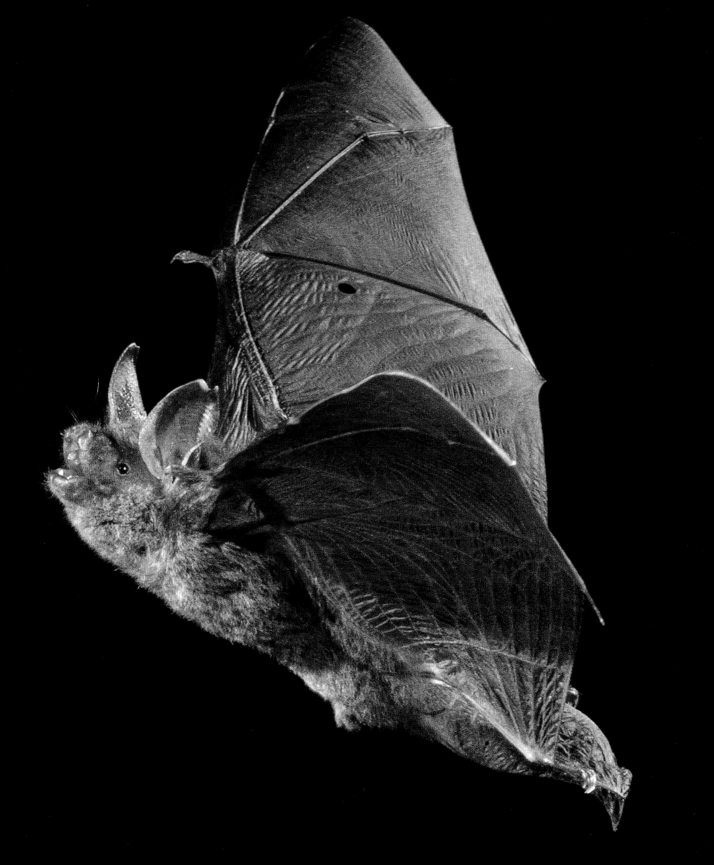

▲ Rafinesque's big-eared bat has a uropatagium about as long as its tail, connected at both ankles.

◀ The common vampire bat has almost no tail and a greatly reduced tail membrane, as seen by the distinct gap between the bat's ankles.

Well, okay, maybe not quite. But this description does reflect the uniqueness of the bat wing and illuminates one of the keys to how bats unlocked the power of flight. The wings of past and present vertebrate animals, from pterosaurs to birds to bats, are all some version of modified arms. In birds, the bones of the wrist, palm and finger have been reduced and covered in muscles and feathers. In bats, the big bones of the arm (humerus, ulna, and radius) have only been slightly reduced while the bones of the fingers have been stretched to the extremes. The bones of these elongated fingers also have more joints than human hands, giving bats superior abilities to make miniscule adjustments to the shape of their wings. Warning to all the bird lovers: this is the part where I try to convince you that bats are better than birds, at least when it comes to flying.

Stretched between those elongated fingers is a swath of thin, elastic membrane. This membrane is not just skin but is also jam-packed full of tiny blood vessels and long, thin muscles. Their highly elastic wings are important for giving bat flight the edge over the flight of birds. The membrane that stretches over the main parts of the wing is called the plagiopatagium, and the small bit of membrane that stretches along the forearm, over the elbow, and to the upper humerus is the propatagium. Between the legs of many bats is another skin membrane called the uropatagium; some groups of bats have this feature, whereas in others it is greatly reduced or completely absent.

# The Physics of Flight

Getting into the air—whether you're a bird, a plane, or a bat—requires going against the forces of gravity. Flying objects and animals use what's called an airfoil to get upward lift. When the wing shape is angled upward, the air above the airfoil moves faster than the wind below it, defying gravity and creating lift. To maintain that lift, animals must flap their airfoils (their wings) to create thrust, keeping that air moving over the thicker top surface. Because bat wing membranes are stretchy, they can generate more lift while also using less energy. As bats move their wings downward (downstroke), it creates an air vortex along the edges of the wings. But when they lift their wings up (upstroke), this air vortex is generated in another location, closer to the wrist joint. This helps give bats greater control over where and how that lift happens, which leads to more maneuverable flight than achieved by birds.

At a fundamental level, flight is a combination of forward movement (thrust) and vertical movement (lift). How much and where these forces are applied is a driving factor for getting into the air and staying there. Bats produce these forces over the course of a wing beat, creating a cyclic pattern of thrust and lift. Bats don't really flap their arms up and down, but instead move in more of a forward scooping motion. This is important because it means that the amount of wing surface area exposed to airflow changes over the course of the flap cycle and can be varied depending on the location and timing of the wing flap during a specific maneuver. A bat's ability to adjust the position of the fingers and the highly elastic membrane of the wings also makes the forces of lift and drag (the force that might push the bat backward) highly dynamic.

The overall geometry of a bat's wing can be very informative about how and where the bat might fly. Aspect ratio is the ratio of a wing's length to its width. Long, narrow wings have high aspect ratios, whereas short, broad wings have low aspect ratios. A wing with a higher aspect ratio experiences less drag and therefore requires less energy in flight. However, this decrease in drag comes at the cost of reduced maneuverability. Having wings with a high aspect ratio might indicate that a bat spends more time flying quickly in open habitats, whereas having wings with a low aspect ratio suggests slower flight and an ability to maneuver through more cluttered areas.

▲ The highly agile and forest-dwelling Percival's trident bat can contort its wings and body in remarkable ways to squeeze through small spaces.

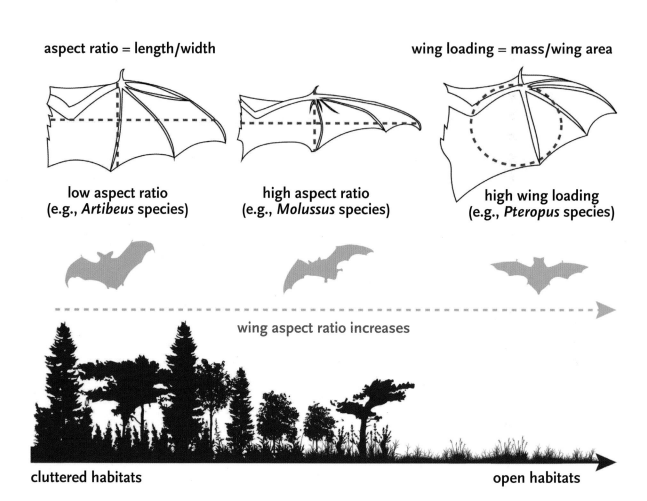

**aspect ratio = length/width**

**wing loading = mass/wing area**

**low aspect ratio**
(e.g., *Artibeus* species)

**high aspect ratio**
(e.g., *Molussus* species)

**high wing loading**
(e.g., *Pteropus* species)

wing aspect ratio increases

cluttered habitats

open habitats

▲ The geometry of a bat's wings can tell us a lot about where and how it flies. Bats with wings of relatively similar length and width (low aspect ratios) are more maneuverable, whereas bats with long narrow wings can fly fast and far over open areas.

Wing loading is another ratio, this time between the mass of the animal (or plane) and the wing surface area. Larger bats have higher wing loading compared to small bats. Increased wing loading means that a bat needs to fly at a faster speed to generate enough lift to stay in the air. Bats with lower wing loading have increased maneuverability compared to bats with high wing loading.

While both wing loading and aspect ratio can be informative in estimating bat flight performance, it is important to remember that these measures were derived from simple, fixed aircraft wings. Flapping flight means that bats can adjust for some of these effects in ways that airplanes cannot, giving them much more flexibility in flight than can be predicted from math alone.

# Fossil Bats

Flight is the primary thing that separates bats from all other mammals. But how and when did they take to the skies? To truly answer these questions, we'd need a time machine to take us back about 60–66 million years ago, to right around the Cretaceous–Paleogene boundary. This boundary is primarily defined by one of Earth's extreme mass extinction events, resulting in the extinction of all nonavian dinosaurs and ushering in the age of mammals. Molecular analyses suggest that ancient bats probably first started evolving around this time. For many animal groups, fossils have helped fill in gaps about where they came from and how they evolved. Unfortunately, for bats it's rare to find good-condition or complete fossil bat skeletons.

According to a recent survey of fossils, the completeness of the bat fossil record is the poorest of any of the four-limbed animal groups. Why aren't there more bat fossils? For one, bats are generally small and small things don't preserve well in the fossil record. Most of the fossils that have been assigned to ancient bats are the more durable parts of the skeleton, such as the teeth, femurs, and the occasional arm bones. Finding the extremely thin and elongated bones that make up the bat wing is incredibly rare.

If it weren't for special formations called lagerstatten, we wouldn't have much of an idea of what the entire skeleton of an ancient bat might have looked like. Lagerstatten stone layers are rich in exceptionally well-preserved fossils of all types of organisms. They are formed when dead plants or animals are buried in environments that limit the ability for bacteria to break down the organic

◄ Black mastiff bats are large and have long, narrow wings. This makes them good at flying fast in open spaces, but they are terrible at flying in tight quarters and cannot take flight from the ground.

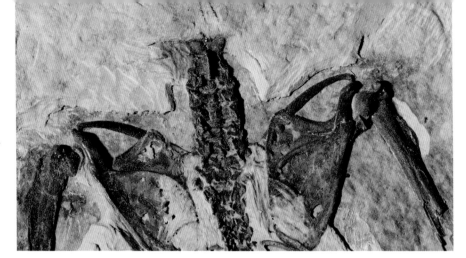

material. This leads to unusually detailed traces of long-dead animals, which can reveal the presence of feathers or even soft tissues—features that aren't normally preserved. Famous lagerstatten include the Burgess Shale in British Columbia and the La Brea Tar Pits in southern California. Another reason it's difficult to find detailed bat fossils is the wooded environments where these ancient bats are thought to have lived, as trees and forests don't usually lead to high-quality fossil sites. However, at least a few lagerstatten have contained complete or nearly complete bat skeletons, giving us some insight into their ancient evolutionary history.

Found in the Green River Formation of Wyoming, Finney's clawed bat (*Onychonycteris finneyi*) is the oldest known complete bat fossil, dating back to the early Eocene (52.5 million years ago). In addition to *Onychonycteris*, many other complete or partial bat skeletons have been dated to this period, found in rock formations around the world. Consistent with morphological data, these fossils suggest that bats had already dispersed and started evolving into many different species by this time. Researchers estimate that there is approximately 10 million years of bat evolutionary history that remains unaccounted for and therefore is still a mystery.

## The protobat

Looking at fossils like *Onychonycteris*, it is almost surprising how much it looks like the contemporary bat. Though folded up, the fossil has clearly delineated wings with the elongated wrist and finger bones characteristic of modern bats, though there are some differences. Most modern bats only have claws at the tip of the thumb, but *Onychonycteris* had a claw at the end of each slender finger. Given the well-developed wings, scientists are certain that *Onychonycteris* was capable of flapping flight, but it probably wasn't the most graceful flier. Using modern imaging techniques, researchers from the American Museum of Natural History digitally reconstructed the wings of two *Onychonycteris* specimens and found that this species has "no modern aerodynamic equivalent." Estimates of wing area and body size indicate that this bat had both a low aspect ratio and very high wing loading—meaning the bat would have to fly very fast to generate enough lift and this flight was probably not very energetically efficient.

When the researchers excluded the "hand-wing" part of the wing from these analyses, they found the values to be a lot like that of modern gliding mammals like flying squirrels. Without the known existence of any transition fossils that would bridge the evolutionary gap from nonflying to flying bat, it's hard to say exactly how bats took to the skies. It seems mostly likely that early protobats were arboreal gliders, floating from tree to tree as they captured insects. One main hypothesis proposes that the webbed arms and hands of these protobats might have even been used to capture insects from the air, not all that different from how some modern bats capture insects in flight. The similarities between *Onychonycteris* and modern gliding mammals helps support this hypothesis of how bats came to dominate the nighttime skies.

◄ The recently described bat fossil *Icaronycteris gunnelli* is thought to be closely related to *Icaronycteris index*, one of the first described fossil bats that was dated to around 52.2 million years ago.

# What came first, flight or echolocation?

Another major unanswered evolutionary question about bats is whether echolocation evolved before or after the ability to fly. Again, we can turn to *Onychonycteris* to help fill in some of the blanks. Comparisons of the structure and shape of the fossil ear bones to those of living bats revealed that *Onychonycteris* more closely resembles non-echolocating bat species. The cochlea, the spiraling inner ear bone, has fewer turns in *Onychonycteris* than in modern echolocating bats. Therefore, although capable of flight, the evidence seems to suggest that *O. finneyi* did not use echolocation. In comparison, *Icaronycteris index*, a slightly more recent bat species also known from fossil deposits in Wyoming, is thought to have used echolocation, based on those same inner ear features.

However, the debate might not be over. Later research raised some doubts about echolocation in *O. finneyi* based on the bones associated with the larynx. In bats that echolocate using their larynx, the ends of one of the throat bones directly connects to the tympanic bone in the ear. Looking at these bones in *O. finneyi*, it appears that a similar articulation exists, raising the possibility of echolocation in this ancient bat. The truth is, we may never know the exact order of these two important milestones in the evolution of bats.

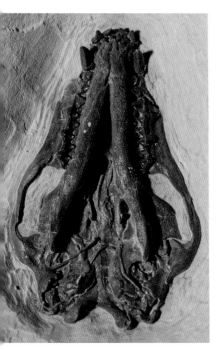

▲ Close examination of the ear and throat regions of fossils like this *Onychonycteris finneyi* skull gives clues to when and how bats evolved the ability to echolocate.

# How to Grow a Wing

The layers of skin and thin muscle that stretch between a bat's fingers are very important for bat flight and one of the things that sets bats apart from other mammals. Comparing the early development of bat limbs to that of other mammals is another way that scientists can start to understand how bat wings (and flight) might have evolved. If we look early enough in the embryonic growth of humans, we also have webbed hands at one point. The main difference is that eventually our webbing goes away, whereas in bats these membranes stick around.

In most mammals, hands first start as limb buds before spreading out into paddle-shaped plates (the process is very similar for feet as well). Gradually, the tissue between what will later become the fingers begins to regress through a process called apoptosis, or programmed cell death. Exactly how this happens is a complicated cascade of different genes turning on and off and the work

of various proteins and enzymes. One notable group of molecules specifically involved in the deterioration of the hand webbing is Bmp, or bone morphogenetic proteins. In bats, in combination with other molecules (examples include Grem and Fgf8), Bmp prevents the degradation of the between-finger tissues during development, leading to the retention of this membrane in fully developed bat fetuses.

The development of membranes between the fingers and ankles in bats is more perplexing, at least from a developmental perspective, as no other animal has a similar structure. The tail membranes of bats are important in giving them additional agility and flexibility in the air and in capturing prey.

## Touch the Sky

We can think of flying as a contact sport, only instead of brushing against another body or tackling another animal to the ground, the contact is between the animal and the air. Along the surface of a bat's wing is a sparse grid of domed, microscopic hairs. Scientists have known about the presence of these hairs for a long time, and the hairs were first proposed as a possible explanation for how bats could fly so well at night—an idea refuted following the discovery of echolocation. So, what are the hairs for?

Mammalian skin is full of cells and nerve endings that convey touch information like pressure, direction, and duration to the spinal cord and brain. For example, Merkel cells are closely associated with free nerve endings under the skin. The microscopic domed hairs on bat wings are mostly associated with Merkel cells, and bats have more of these hair–Merkel cell associations than other mammals (about 50 percent of the hairs in bats compared to only about

► With their long, narrow wings, velvety free-tailed bats need to gain some height to take flight.

▲ A big brown bat propels itself off a piece of wood. When taking off from horizontal surfaces, most bats can push themselves up using their forearms.

2 percent in mice). However, these hair–Merkel cell combos are not evenly distributed over the surface of the wing, with most found along the fingers of the bat. When scientists removed these hairs using depilatory cream (which breaks down the base of the hair and is commonly used as a beauty product in humans), it also changed the way bats flew. Bats navigating through an artificial forest moved faster but were also less maneuverable than bats that had intact wing hairs. So, something about those hairs is important for helping give bats their extraordinary acrobatic flight talents.

To try to understand exactly what these hairs might be doing to help bats, researchers puffed tiny amounts of air over the hairs and recorded neuronal responses in the brain of the bat. They found that while these hairs and Merkel cells were not sensitive to the duration or strength of these air puffs, they responded differently to air flowing from different directions. Specifically, the nerves had the highest response to reverse airflow or air that moved from the trailing part of the wing toward the leading, top edge of the wing.

When bats are flying, air moves over the hairs, causing them to bend in certain directions. This change in shape is detected by nerve endings and Merkel cells, giving the bat information about its speed. If a bat is flying too slowly, it is at risk of stalling (or falling out of the sky). When the wing hairs were removed, bats no longer had a way to reliably gauge airflow patterns across the wings, so they probably felt like they were at risk of stalling, causing them to speed up in compensation. Combined with other touch receptors embedded in the membrane, these hairs give bats a lot of feedback on the position of their wings during flight, which gives them the advantage they need to complete the stunning flips, twists, and turns they execute in flight.

# Bats on the Ground

The hand-wing of the bat gives it a lot of advantages in the air. But any good flying animal needs to be able to manage on the ground at least a little bit, for those times when flight goes wrong. How well bats can move on the ground—and get themselves back into the air—depends on their body and wing morphology. Although some bat species cannot take off from the ground, plenty of species are able to launch themselves into the air from the ground. For

example, many Neotropical fruit bats can spring into the air from a belly flop, taking advantage of their relatively broad wings and stretchy tendons in their upper arms.

For bats that don't have the right anatomy to spring themselves off the ground, they need another strategy for when they accidentally end up on the ground. Bats that have wings with high aspect ratios (long narrow wings) need more time to get enough lift beneath their wings before they can catch enough air to fly. While it is far from graceful, narrow-winged bats like the velvety free-tailed bat can scurry along the ground at least long enough to find a tree or other vertical structure that gives them the height they need to take flight.

Two species of bats deserve special recognition for their ability to ambu-late on the ground: the New Zealand lesser short-tailed bat and the common vampire bat. These bats independently evolved the ability to move well on the ground, meaning they don't share a walking bat ancestor. To study how these bats move on the ground, researchers trained them to walk and run on a treadmill while recording them using high-speed video. Both bat species were observed to use a lateral sequence walk, a widespread form of walking in ter-restrial animals where the left forelimb moves forward at the same time as the right hindlimb. Common vampire bats were notably very smooth during this movement, keeping their body at a constant height to move with a smooth, horizontal, catlike gait.

As the treadmill speed increased, common vampire bats would shift from this smooth walk to a bounding gait, complete with a phrase of dramatically sailing through the air. While this bounding gait is kinematically unlike the walking and running gaits of other animals, it can be defined as a run because all four limbs are off the ground during at least one phase of the movement. The run is initiated by the bat pushing off its forelimbs (wings), followed by the hindlimbs. Using this gait, vampire bats can run as fast as 2.5 miles per hour, a pretty good pace for an animal that only weighs about 30 g! However, they don't seem able to maintain these high speeds for long and probably only use this fast-moving gait to get themselves out of danger—like avoiding the hooves of their 2000-pound prey.

New Zealand lesser short-tailed bats don't have the sprint speed that com-mon vampire bats do. In contrast to the blood feeders, the New Zealand lesser short-tailed bat is an omnivore, feeding both on insects and plant material.

While these bats can capture insects from the air, they spend up to 40 percent of their hunting efforts on the ground, digging through leaf litter or feasting from plants, making them the most terrestrial bat in the world. This bat is a particularly important pollinator for the endangered wood rose (*Dactylanthus taylorii*), a ground-growing plant that is endemic to the islands of New Zealand.

To accommodate all this time on the ground, the New Zealand lesser short-tailed bat has a variety of adaptations related to ground movement, including small talons at the base of the thumb and toes, grooves in its foot pads to increase grip, and a special sheathlike portion of the wing that protects those small, delicate bones during walking. Exactly why these bats evolved the ability to move so well on the ground is still uncertain. One idea is that the lack of ground-based predators on the islands facilitated the evolution of walking and burrowing in these bats, like the way flightlessness evolved in many island birds. Fossil evidence from the extinct Australian bat *Icarops aenae*, an ancient relative of today's New Zealand bat, suggests that at least some bats could be found hunting on ancient forest floors. As these bats expanded into what later became the islands of New Zealand, the lack of predators meant they could take advantage of the variety of plants and insects available on the ground.

▼ To study the walking gait of vampire bats, researchers trained them to walk on a small treadmill.

▲ A common vampire bat perches on a rock inside a cave entrance, revealing the extra-long thumbs that help this species move well on all four limbs.

## Chapter 7

# SHARE

Animal groups are everywhere: the mesmerizing shift and shimmer of a huge school of fish moving as though they are one; wildebeest, zebra, and antelope blanketing the Tanzanian plains, as far as the eye can see. But if you've ever had to squeeze through an uncomfortably crowded sidewalk or been the last to the buffet table at a party, it can seem like group living has its downsides.

So why do animals form groups? Large groups can make it less likely that a specific individual might get eaten by a predator. For example, cooperative swimming of schooling fish makes it harder for a dolphin to capture an individual fish, and a fast-moving stampede of wildebeest challenges lions looking for easy prey. Cooperation between individuals can mean access to more energy-rich food, extra time to eat while the individual next to you keeps an eye out for predators, maintenance of body temperature in the cold through huddling, and a variety of other benefits. While these behaviors might benefit the group, they are still mostly the result of individuals looking out for themselves.

▶ Thousands of Mexican free-tailed bats stream out of Bracken Cave in central Texas. This cave is the world's largest known bat roost, home to between 15 and 20 million bats each summer.

True selflessness or altruism—behaving in ways that benefit another individual at the cost of oneself—is not very common in animals other than humans. Despite their gruesome reputation, vampire bats are a fascinating group of animals to study to better understand cooperation and altruism, how these behaviors might have evolved in humans, and even to learn more about friendship.

# Generous Vampires

Blood is not the most substantial or nutritious meal. Vampire bats need to feed almost every night to survive, risking starvation if they miss more than two consecutive nights of hunting. Luckily, common vampire bats are willing to share, by regurgitating stored blood into the mouth of another bat. Getting this mouth-to-mouth blood infusion can help keep a not-so-successful vampire bat alive to hunt another night. But why would an individual sacrifice its own night of hard-earned blood for another bat?

In the late 1970s and early 1980s, Gerald Wilkinson and his collaborators spent hours staring up from the floors of vampire bat roosts to try to answer that question. Wilkinson observed that within a roost, female vampire bats seemed to prefer roosting, grooming, and sharing blood with certain individuals over others. They first thought that these preferences were based on familial relations, with females choosing to share with their sisters or mothers. Since an individual shares some percent of genes with their relative, by helping that relative to survive and reproduce, even at some cost to themselves, the individual's genes will continue to persist in the population. This type of evolutionary selection is referred to as kin selection. Wilkinson and his colleagues did find that donor and receiver pairs were often related to each other, but relatedness wasn't the whole story. Past associations between individuals were also important. Bats were more likely to regurgitate blood to individuals who shared blood with them in the past, forming a reciprocal relationship.

Later work by one of Wilkinson's graduate students, Gerald Carter, supports this conclusion. Carter found that receiving food from another vampire bat in the past was the best predictor of whether a bat would donate a blood meal. Interestingly, they observed that blood meal recipients rarely begged for food, and it was the donors who usually initiated food-sharing events. Expanding

food-sharing networks beyond relatives helps reduce the risk of starvation. Termed social bet-hedging, by investing in many relationships bats can protect themselves from an unpredictable environment or loss of a current social partner. After experimentally removing a key food-sharing partner, researchers found that females who fed a greater number of unrelated individuals suffered a smaller reduction in food received than females with smaller social circles.

Vampire bats might be willing to share food, but they don't dive into a food-sharing relationship with just anyone. In the wild, common vampire bats form relatively stable social groups, meaning the same individuals roost together over months, even years. Bats in these small groups groom each other frequently, and this grooming behavior is probably important for both maintaining and developing food-sharing and cooperative bonds. Compared to other bat species, vampire bats spend much more time grooming each other. Research by both Wilkinson and Carter indicated that there is a relationship

▲ A colony of common vampire bats roost together inside a hollow tree in Central America.

SHARE

▲ A common vampire bat leaves the roost for a night of foraging.

between social grooming and food sharing, with bats that groom each other often being more likely to also share blood meals.

More recently, Carter and his colleagues investigated how these relationships might form between bats. They put wild-caught and captive-bred females together in the lab and then performed fasting trials, with one individual isolated without food for 24 hours. They found that bats preferred grooming and food sharing with their original roost mates. When introduced in pairs with unfamiliar individuals, these bat strangers would eventually start grooming each other. The rate at which bats groomed unfamiliar partners predicted if they later went on to become new food-sharing partners. By starting with a low-investment behavior like grooming, bats can gradually increase their commitment to the relationship and reduce the chance of being exploited by a cheater—a bat who takes but never gives.

Our understanding of food-sharing, social grooming, and other cooperative behaviors between individual vampire bats has been mostly limited to what goes on in the roost. Although the roost is undoubtedly where bats spend a large majority of their time, what about when bats are out hunting? More recently, Simon Ripperger and Gerald Carter investigated how these inside-roost relationships influence what goes on outside the roost (or vice versa). Using the same combination of wild-caught and captive-bred bats, they released the bats into the wild with some new accessories. Attached to each bat was a small computer sensor that emits a signal every 2 seconds and can detect when it is near another sensor. These proximity loggers record when two individuals are within a certain range (in this case, between 5 and 10 m) of each other, roughly how close the bats are (using signal strength), and how long the respective bats stay close to each other. They found that cooperative behaviors recorded in captivity (social grooming and food sharing) were predictive of how much time bats spent together outside of the roost while foraging. Even though bats left the roosts separately, they frequently reunited while hunting. During this study, Ripperger and Carter also recorded video and vocalizations of bats while out foraging. They observed bats making downward sweeping calls and a potentially newly described n-shaped call, which is thought to be a foraging-specific call.

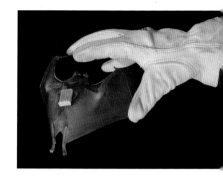

▲ A common vampire bat is released after being fitted with a small proximity logger.

Another species, the white-winged vampire bat, will also exchange calls in a duet-like fashion in circumstances when individuals are separated, such as when leaving the roost or while feeding. The calls are a little different than those of the common vampire bat, forming downward sweeping notes emitted in series of two, but they contain enough variation in the call structure for bats to recognize the contact calls of specific individuals. These calls help bats of a given social group reunite at a feeding site or find each other to share a blood meal. Future work is still needed to understand how inside-roost relationships might extend to other social situations in the lives of vampire bats.

# Bat Chat

If you have ever stood near an active bat roost in the summer, you might notice that bats can be pretty chatty—lots of little squeaks, chirps, peeps, and trills. Just like other animals, bats use sound to communicate. Separate from echolocation calls, which are used for navigating, these sounds are collectively referred to as social calls. Some of these calls are within humans' hearing range, while others can match the ultrasonic frequencies of echolocation pulses. Bats grumble at each other as they fight over food or roosting space and screech to defend themselves or warn nearby bats. They also call when trying to attract mates (more on that later). These social calls can also be a way for bats to coordinate among group mates to share resources like roosts and food.

# Call and response

Spending hot summer days at the local community pool, my sister and I played many a game of Marco Polo. This modified form of tag involves one person keeping their eyes closed and periodically calling out "Marco," to which the other player(s) respond "Polo." The game ends when the person calling "Marco" finds the other player(s), using only sound cues.

Some bats show a similar call-and-response behavior when moving through their forest habitats. Only instead of trying to avoid the questing individual, like in the childhood game, bats use call and response to recruit others to join them at a roost or a food source. When your preferred roost spot is only available for about 6 to 30 hours, having some help in finding the next sleeping spot can be helpful.

Spix's disc-winged bats, which roost upright in the furled leaves of *Heliconia* plants, play their own call-and-response game in the wild. Young *Heliconia* leaves start out rolled up like a newspaper before eventually spreading out, a process that may take less than a day. As a result, disc-winged bats need to be able to find a new furled-leaf roost almost every day. Gloriana Chaverri from the University of Costa Rica and Maria Sagot from the State University of New York at Oswego have spent many years investigating how these little bats use social calls to coordinate in the forest. Flying disc-winged bats produce inquiry calls as they roam the forest, looking for their next sleeping spot. Bats that have already found a cozy leaf will emit response calls in answer to these inquiry calls, thus helping the flying bat find a new roost spot—and a roosting buddy. The shape of the leaves themselves also act like acoustic horns, amplifying the sound as bats call from within the roost and making it easier for the flying bats to hear. This bat species forms small social groups of around five or six individuals, who consistently try to share roosts throughout the year. Flying bats can distinguish the response calls of individuals and usually choose to roost with familiar individuals over strangers.

# Making contact

Using sound to locate a roost mate or another individual of the same species is a common behavior among animals. Although most bats don't use the more structured call-and-response behavior described above, contact calls are

▲ Spix's disc-winged bats call to each other to coordinate meetups in their rolled-up leaf roosts.

used by bats in a variety of social interactions. Although the exact characteristics of these contact calls depend on the species making them, they are lower frequency than echolocation calls, with the sound frequency sweeping from a higher pitch to a lower pitch.

Contact calls are used by many species to find each other at roost sites. Several species of tent-making bats have been observed emitting calls at or near roosts. Thomas's fruit-eating bat males call from under their leaf tents, possibly to recruit nearby females to come roost with them—like the bat version of a "Netflix and chill" invitation. Honduran white tent-making bats also make social calls near potential roosts, maybe also trying to alert nearby individuals about a favorable roosting resource. Pallid bats frequently switch roosts, moving to new rock crevice spots as often as every 1–2 days. Both after leaving the roost and during their return, pallid bats repeatedly fly back and forth near the roost while producing multiple loud calls in rapid succession. Recording and playback data from pallid bat colonies in Oregon suggest that these calls serve to maintain contact with each other at the beginning and end of a night of foraging.

▲ Small tent-making bats like these may use contact calls to recruit conspecifics to their roosts.

SHARE

► Great fruit-eating bats some-times emit grating, high-pitched calls when being handled. These calls appear to be attractive, as I have fre-quently experienced other bats swooping toward me while handling a crying bat.

# Help, help! I'm in distress!

Many animals—from frogs to songbirds to primates—will make certain types of noises when in distress. These calls tend to be loud and noisy calls, a type of grating shriek that seems to rake itself over your ears and through your body. In birds, distress calls can serve a range of functions, including startling a predator into accidental release or warning, repelling, or attracting fellow birds over to help with the fight (referred to as mobbing behavior).

Not only do bats also produce these large-bandwidth distress calls, they are often attracted to the sound of these calls. Foraging least horseshoe bats respond to distress call playback by repeatedly swooping over and near the source of the sound. Other studies have found that playing distress calls near mist nets can increase captures, as bats seem to come down and check out the source of the noise. But why would an individual get close to a potentially dangerous situation?

One hypothesis is that bat distress calls might promote cooperative mobbing, a way to attack or drive off a potential predator. In one study, researchers broadcast the distress calls of the velvety free-tailed bat, a fast-flying but not very maneuverable species. Although velvety free-tailed bats made more passes over the speaker, the bats did not exhibit the group approach or attack behaviors associated with mobbing in other animals like birds. In another study, researchers found that sac-winged bats would approach distress calls in the roost but not when near foraging sites, perhaps because a potential predator near a roost is a bigger danger than one near a foraging spot. Current evidence suggests that it is unlikely bats are engaging in cooperative behaviors by responding to distress calls. Instead, they are more likely to be eavesdropping on the call of conspecifics to evaluate their own risk.

## The Potato Chip Effect

Sharing the search for food and then sticking around to eat in groups can be advantageous. The more individuals out looking for the same resource, the more likely one of them is going to find something good and easily sharable. For bats, this could be a large fruiting tree with more than enough fruit for all the roost mates or a dense swarm of insects. Staying in a group can also make it easier to defend food or other resources from other interloping bats or minimize the chance that one specific individual might get nabbed by a predator.

Sharing social information can be especially helpful for bats that rely on ephemeral food sources, like insect swarms that might be in one place for only one night. In the case of the Mexican fish-eating bat, this temporary food source comes from short-term ocean upwellings of fish and crustaceans. Using tiny GPS tags and attached ultrasonic microphones, a group of international researchers tracked the nightly movements of a few fish-eating bats. By listening in on the bats' environment as they traveled between roost and sea, they found that individual bats traveled with at least one other fellow fish-eating bat for much of the night.

To test if the fish-eating bats were taking advantage of the calls of their neighbors, the researchers performed a playback experiment. They broadcasted three types of calls: fish-eating bat echolocation calls, bat calls with a feeding buzz (indicative of active feeding), and white noise as a control. Fish-eating bats were five times more likely to approach the sound of a fellow bat's echolo-cation call and fifteen times more likely to approach a feeding buzz call when compared to the white noise control. In this case, bats are not specifically pro-ducing calls to attract their fellow conspecifics, but just eavesdropping on each other's sounds. This type of eavesdropping is sometimes referred to the "potato chip effect." Imagine you're in a dark movie theater and someone opens a bag of chips. Everyone is going to hear that noise and know roughly where that person is. By listening to the hunting sounds of those nearby, bats can improve their own ability to track down food.

Unlike fishing bats, some bats actively try to recruit other bats to join them while hunting. Greater spear-nosed bats are large omnivorous bats of the Neo-tropics that are frequently observed foraging for food with their colony mates. The bats fly long distances to reach foraging sites; in one study in Panama, indi-viduals flew across about 25 km of water to access flowering balsa trees. They mostly made these trips alone, later meeting up with colony mates at the flow-ering trees. Greater spear-nosed bats often emit loud screeching vocalizations both while eating and outside of roosts. These calls are made more frequently when bats are flying near each other, and they seem to serve as acoustic sign-posts for bats to recruit others to a food source and help them reunite at feeding sites following solitary commutes. Colony mates have screech calls that sound more like each other than to the calls of another social group, helping bats keep track of their own hunting cliques.

Emitting social calls at foraging grounds is one way bats may share the loca-tion of food, but roosts can also be important places to exchange information about potential food sources. Captive short-tailed fruit bats were able to learn about new food options by watching and interacting with demonstrator bats, who had been previously trained to feed on new foods. Peter's tent-making bats show similar abilities to learn from roost mates about new foods and are able to distinguish between different cues to learn the quality of these new resources. Naïve bats learned to prefer a new food by smelling the new food odor on the roost mate's breath, but they didn't show a preference when that new smell was simply on the roost mate's fur. Sharing this information might not be a con-scious choice. By having fruity breath, however, bats may help fellow bats learn about the presence of certain foods or expose them to new food resources.

◄ Tent-making bats can learn about a new food resource by smelling it on the breath of their roost mates.

# Frenemies

What happens when bats don't want to share? Mexican free-tailed bats, famous for living in extremely large colonies that can number in the millions, would sometimes rather keep that tasty insect for themselves. At foraging sites in the southwestern United States, Mexican free-tailed bats were recorded emitting an unusual social call. When visualized by plotting the frequency of the call over time, it looks a bit like a wonky scribble of a snake, a squiggly up-and-down pattern termed a sinusoidal frequency-modulated call, or sinFM for short. Bats only produce sinFM calls in the presence of other free-tailed bats and only when that other bat is making a feeding buzz indicating it's about to chomp down on an insect.

In one experiment, hunting bats were up to 85 percent less likely to capture an insect in the presence of an sinFM call, suggesting that the calling bat is actively thwarting the other's chance of making a catch. To work, the call must be emitted at the same time as the other bat's feeding buzz. When researchers broadcast sinFM calls just before the start of the feeding buzz, it did not prevent insect capture. This bat-jamming behavior isn't completely petty, though. Based on the flight trajectories of bats as they were making these sinFM calls, the calling bats were simply trying to use all the advantages they could to get to a particular prey item first.

# Sharing Airspace

The summer sun is still shining bright, though the rays of light are shifting to the warm, burnished tones of golden hour as sunset approaches. I'm sitting on a wooden bench in a gravel viewing area. Just down the hill, nestled in a mottled cliff face covered in brown and sage green grass and patches of prickly pear cacti is the oblong opening to Bracken Cave. A hush begins to fall over the small crowd of people who have also gathered to watch the emergence of nearly 20 million Mexican free-tailed bats as they get ready to head out over the central Texas skies for a night of hunting.

Flashes of movement are just starting to be visible in the dark maw of the cave, as the bats begin circling in preparation for their exit. Slowly, a ribbon of bats begins to stream out of the cave. As we watch, this little trickle of bats gradually grows until it's a swirling, circling river and we are engulfed in the sounds of many tiny wings pushing against the air. At any given moment, thousands of bats may be vying for the same airspace, all producing echolocation calls as they navigate themselves out of the cave. If we could hear ultrasonic frequencies, the sounds of flapping wings would be accompanied by a cacophony of bat calls. How are bats able to orient themselves in this mishmash, keep track of their own call echoes, or even hear over the noise?

Bat scientists have pondered these questions for a long time. Given how important the acoustic world is to bats, it's not surprising that they have evolved a variety of strategies that help them manage this complex acoustic puzzle. One potential strategy that bats might use to reduce this cocktail party nightmare is the jamming avoidance response. First described in weakly electric fish, the jamming avoidance response occurs when a bat reflexively adjusts its echolocation call in a way that minimizes that call's similarity to the calls of another bat. One type of adjustment is changing the pitch or sound frequency to better distinguish its call from that of a fellow bat. When flying Mexican free-tailed bats were presented with recorded conspecific calls, they responded by shifting their call frequency upward, specifically at the end of each call. The European free-tailed bat uses a similar strategy, though both species will shift their calls both up and down in response to other bat calls. Even with the small frequency changes, however, there's still a lot of call overlap between individuals of the same species.

When humans are talking in a noisy area, we involuntarily raise the loudness of our voice to separate our words from the background. This phenomenon is called the Lombard effect and is known to occur in birds and mammals, including bats. Changes in call loudness have been observed in greater horseshoe bats, but only when the other sound overlapped with the dominant frequency component of a calling bat. Kuhl's pipistrelle changes its echolocation calls by emitting both higher intensity and longer calls in situations that simulate extreme interference, like that of a high density of other bats.

Bats will also adjust how much and when they call in response to interference. Big brown bats flying in pairs in the lab were observed to fall silent for some period, with at least one bat stopping vocalizations for more than 200 milliseconds. Bats were more likely to fall silent when the other bat was flying close (less than 0.5 m) and when bats were flying toward each other. While these periods of silence weren't long, they might be just long enough to help modulate the potentially interfering echoes from other bats. In a similar

▲ Thousands of Mexican free-tailed bats emerge from Bracken Cave in San Antonio, Texas, on a summer evening.

strategy, Mexican free-tailed bats will adjust their calling rates in response to the playback of other bats. When flying in a flight room by themselves, free-tailed bats emit echolocation pulses in a steady, even rhythm. When exposed to a playback simulating another bat, they reflexively drop a pulse from their pattern, resulting in about a 15 percent reduction in call rate compared to when flying alone. This mutual suppression can result in the appearance of a cooperative behavior. By dropping a pulse here and there, bats are unlikely to be losing that much information from their environment but gain the benefit of improved navigation when another bat of that species responds the same way.

With so much variability in echolocation calls both within and between bat species, it is difficult to make sweeping conclusions about how bats might be dealing with the cocktail party nightmare. Bats might also use one strategy to avoid interference from competitors while hunting, but a different one when just flying straight in a group. In watching the spectacular emergence of thousands of bats like those at Bracken Cave, it's also easy to imagine graceful movements as bats flow into the sky. While the sight is mesmerizing, it's not like the highly coordinated murmuration of starlings forming shapes across the sky or schools of fish. Bat emergence flights are a little more like trying to get through the front door of a big box store during a Black Friday sale—mostly polite, but still some contact and maybe the occasional elbow to the arm. Bats frequently run into each other, clipping wings or the edges of the cave entrance as they exit. While the previously described strategies might help in this period of intense activity, it may be that bats just have to suffer through until they can get themselves somewhere more isolated.

▶ A slightly less crowded moment as Mexican free-tailed bats emerge from Bracken Cave.

# HOME

## Chapter 8

Home is where you hang your bat. Well, maybe that's not *quite* how the saying goes, but roosting habitat is incredibly important in the lives of bats.

Roosts are where bats rest during the day, perform courtship rituals, mate, hide from predators, give birth to and nurse young, and spend the winter in hibernation. The types of roosts chosen by bats are as varied as the bats themselves. Some stay loyal to the roost where they were born, returning every year to give birth, whereas other species might switch roosts every few days. Bats might rely on several roosting spots throughout the day, using night roosts to handle prey and take a break during foraging and day roosts for resting and socializing. Roost preferences can also change seasonally, as resident bats seek thermal environments to conserve energy during hibernation, whereas other bats migrate long distances to warmer climates during the winter.

▶ A Thomas's fruit-eating bat hangs from a tree branch.

► A short-tailed fruit bat hangs by one foot beneath a bird-watching platform in Panama.

# Let's Hang!

Bats spend most of their lives upside down, a behavior that helps facilitate a rapid takeoff into flight. In addition to the extreme lengthening of the hands to form wings, bats also have knees and feet that face the opposite direction of most mammals. The hind limbs of a bat are positioned so the knees bend pointing backward, and the bottom of the feet face forward. While this configuration has its advantages for flight and prey capture, it makes moving on and getting off the ground difficult (although not impossible). As bats evolved flight, there were selective pressures to lessen the load, leading to lightweight and delicate bones. This means that bat femurs cannot withstand the type of compression stress that would come from upright standing or perching. Suspending their weight from above instead of standing reduces the stress on these bones. Finally, an ability to hang combined with flight might have also opened unique hiding or roosting places that predators and competitors couldn't access—like the ceilings of caves or vertical faces of rock cliffs.

Bats also have unusual features in their ankles, feet, and toes that make upside-down life all the easier. In humans, when we grab something, whether it's carrying groceries or rock climbing, we expend energy contracting the flexor muscles in our hands. Eventually we get tired and lose our grip strength. However, bats have a tendon-locking mechanism that makes it possible for them to hang upside down with minimal muscular effort. The parts of their ankle tendons that connect the toe flexor muscles to the bone have bumpy tubercules along the surface. The tendons themselves are enclosed in a tunnel-shaped sheath, which has ridges along its inside. When the flexor muscles contract, the tendon is pulled through this tunnel and those bumpy areas catch on the ribs of the sheath, locked in place by the bat's body weight. This locking mechanism is so effective that even in death, everything stays locked in place, leaving dead bats hanging.

# Life Right Side Up

While most bat species roost hanging from their toes, a small handful of bat species have taken a different approach to roosting. Bats in the families Thyropteridae (disc-winged bats) and Myzopodidae (sucker-footed bats) instead cling to the inside of leaves by the wrists, with the top of their heads oriented toward the sky. These two families inhabit opposites sides of the world, with disc-winged bats living in the Neotropical forests of Central and South America and sucker-footed bats found only on the island of Madagascar, off the coast of Africa. Although both groups of bats share a similar roosting arrangement, the way they do it is different.

Five species of Neotropical disc-winged bats (genus *Thyroptera*) have been described. These small bats, weighing about 4 g, feed on insects and roost inside furled palm leaves. Disc-winged bats have small, concave discs at the base of their thumbs and hind feet. While the discs themselves have no muscles within, bats can manipulate the shape of the underlying cartilage using their thumbs or toes. This creates a suction-cup effect, allowing the bats to cling to smooth leaves. They also supplement suction with another mechanism known as wet adhesion, in which two solids are held together by a middle layer of liquid. Wet adhesion does not require energy and so may be a way for disc-winged bats to remain attached to the roost surface without having to constantly contract their thumb or toe muscles.

Over in Madagascar, the two species of sucker-footed bats (genus *Myzopoda*) also have small pads on their wrists and ankles, allowing them to roost head-up while clinging to leaves. Despite the name, however, sucker-footed bats don't use suction to stick to the sides of leaves. Their adhesive pads are flat or convex, with

no cartilaginous plate, and stick to surfaces only using wet adhesion. Researchers noted that when the wrist pads were pushed forward, they unpeeled easily from the surface. This orientation probably helps the bats easily detach and reattach as they crawl over the surface of the leaves, but it also means that roosting upside down would be near impossible because gravity would pull forward on the pads and cause passive detachment.

So far, these seven species of bats are the only ones known to regularly roost facing upward. Despite living and evolving on opposite sides of the world, these two groups of bats converged on the habit of roosting in furled leaves that open at the top, which resulted in fairly similar mechanisms for sticking to those leaf surfaces. Roosting upright in these leaves probably also facilitates a fast escape during disturbance, as the bats don't have to back themselves out of the leaf funnels to get away from potential predators.

# To the Bat Cave

Either through popular media like Batman cartoons or personal experiences, most people associate bats with caves. In North America, almost half of all bats rely on caves for roosting during at least some part of the year. In other parts of the world, cave use might even be higher, with an estimated 77 percent of China's bat fauna thought to roost exclusively in caves and other underground habitats.

Roosting in caves offers many advantages. They are dark places where bats can wait out the daytime, safe from disturbance and predators. Caves are also permanent and available year-round, meaning bats don't have to find a new place to roost each night or each year. Caves are generally well insulated from outside climate conditions, making them good places to rest for small animals like bats that need to regulate their body temperatures. Some bats prefer hot and humid caves in tropical regions. In temperate areas, however, bats seem to prefer stable, cool environments, especially important during winter's freezing cold temperatures and precipitation.

One major disadvantage of roosting in caves is that, although they might be permanent and potentially very large structures, the distribution of caves in the landscape can be highly variable. The formation of caves—one of the most well-known features of karst landscapes, which also include underground lakes

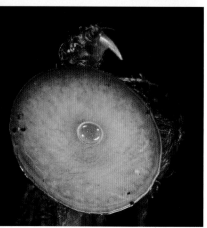

▲ Instead of hanging upside down, Spix's disc-winged bats roost head up inside rolled-up leaves.

▲ Spix's disc-winged bats cling to the smooth leaf surface with the help of disc-shaped structures on their wrists and ankles.

and streams, sinkholes, and springs—requires certain conditions. Karst and cave habitat is associated with rock types like limestone, marble, and gypsum. These rocks are water soluble, so as rain and other water sources move across the bedrock, it creates cracks and fissures that expand into large cavities and caves. The appropriate rock formations can be geographically patchy, with some parts of the world more likely to contain caves than others. Caves might be found in high densities in areas with a lot of limestone or calcium carbonate, such as Southeast Asia, whereas areas with denser rock are less likely to contain caves or caves large enough to house large numbers of bats.

# Up in the Trees

Many bats across the world rely on caves for safe shelter, but just as many of the world's bats rely on trees and other plants for their roosting needs. Though tree and plant roosts might not be as permanent or stable as cave environments, they are much more abundant, meaning that anywhere trees can grow, bats can find refuge.

Tree cavities are widely used tree-based roosts. These cavities can exist in a diverse range of shapes, sizes, and locations on a tree and are formed in both living and dead trees. Many vespertilionid bats use tree cavities, particularly in temperate regions. Neotropical leaf-nosed bats and sac-winged bats also roost in tree cavities in the Central and South America.

In some living old-growth trees, basal cavities can form in the interior, usually because of exposure to fire. Some of these cavities are almost cavelike, offering bats stable temperature and humidity, protection from rain, space for light to enter, and a long-lasting roost site limited only by the life span of the tree.

Dead trees and snags are also important resources for cavity-roosting bats. Hole or cavity formation can take a long time in trees, and these spaces are more common in older and dead trees. Even fallen trees can continue to be used as roosts by some bats, including several species of Neotropical fruit- and insect-eating bats.

Choice of roost by bats can depend on a variety of factors, including cavity availability, thermal environment, roost size, ease of access, and protection from predators. In general, it seems that most cavity-roosting bats don't have preferences for certain tree species but select cavities based on other physical

▲ Horseshoe bats space themselves among the cracks and crevices of a cave in Thailand.

▲ A black mastiff bat returns to its tree cavity roost in Lamanai, Belize.

▲ Proboscis bats line up on the trunk of a guanacaste tree in Belize. The bats' mottled and shaggy fur and markings on the forearms help camouflage them against the bark.

◀ An eastern red bat day-roosts in a pine tree. Note how this species can pull its furry tail membrane over its body like a little cozy blanket.

characteristics. For many species of temperate bats, larger trees are preferred since they provide better insulation and less temperature variability over the course of the day. Dead trees might have larger and more abundant cavities, but they are also less well insulated than living trees. Where in the forest a given tree is located also matters. Some bat species prefer to roost near water sources or forest edges or in areas with short understories—preferences that are likely related to aspects of the species' biology, such as diet and maneuverability in flight.

Holes and cavities aren't the only places that bats can find cover in trees. As trees age or decay, their bark can start to peel away from the trunk, and the space beneath this exfoliating bark can provide alternative roosting space for bats. Some bats don't even bother hiding, instead roosting on relatively exposed areas of the sides of trees or branches. For example, the proboscis bat roosts in groups as large as forty-five individuals, lined up vertically along the side of a tree. Their shaggy fur and mottled coloration help the bats remain inconspicuous against lichen-covered bark.

The leafy canopies and sprawling branches of trees also serve as important roosting habitat for bats. Flying foxes and other pteropodid bats in Asia, Africa, and Australia can be seen hanging from the tops of canopies in large, noisy aggregations called camps. Rather than hunkering down in cavities or beneath bark, these large bats hang directly from tree branches, sometimes obscured by foliage.

Smaller bat species around the world can also be found roosting against dead and living foliage. Bats that roost hidden among unmodified hanging

▲ Wahlberg's epauletted fruit
bats roost in a tree in Kruger
National Park, South Africa.

leaves tend to be more solitary, roosting either alone or in small groups (usually consisting of a mother and her pups). Bats in the genus *Lasiurus* are prime examples of solitary foliage-roosting bats: the eastern and western red bats prefer to roost in hardwood trees like cottonwoods, whereas hoary bats are commonly found roosting in evergreen trees including Douglas fir and towering redwoods.

The downward facing skirts at the base of palm and banana trees are also common roosting spots for bats in subtropical and tropical areas. Northern yellow bats like to roost in the dead, hanging fronds around the base of palm trees, whereas Africa's banana serotine roosts in the rolled-up leaves of banana trees. The brightly colored painted bat of Southeast Asia also frequently roosts among banana leaves—although this species has also been found roosting in tall grasses, abandoned bird nests, and even in a conical piece of cloth attached to a wire that marked the boundary of a rice paddy field.

Although some bats just roost in leaves, others make leaves into a house. At least nineteen species of bats are known to modify leaves by chewing along the veins in different ways to create leaf tents. Most tent-making bats are in the family Phyllostomidae, though at least three species of Old World fruit bats (Pteropodidae) also modify leaves into tents. These leaf tents can take on different architectural styles, depending on the size and shape of the leaf and how the

HOME

bats chew. Large, fan shaped leaves are usually chewed in circular or semicircular patterns, whereas oblong leaves are cut parallel to the main stem to form a boat-shaped tent. Bats then hang along the creases created by the cuts. Depending on the complexity of the cuts needed to create a tent, the life span of these roosts and the time it takes to create them can vary. Male greater short-nosed fruit bats will spend months chewing, cutting, and creasing the leaves and stems of plants into a roost tent, an effort that helps attract females.

# Weird Bat Roosts

Some bats just do things their own way and have evolved highly specific roosting preferences. The upright-roosting disc-winged and sucker-footed bats are one example of extreme roost specialists, but they certainly aren't the only ones.

The golden-tipped bat is an uncommon bat that lives along the eastern coast of Australia and parts of Papua New Guinea, where they prefer to roost in the bottom of bird nests. They don't roost in the same compartment as the birds, instead chewing out their own basement apartment, where their woolly, golden fur acts as great camouflage against the mossy and twiggy bird nest.

Over in the tropical forests of Panama, another bat uses its jaws to build the perfect home. The white-throated round-eared bat excavates and maintains roost cavities in active termite nests—and the termites don't seem to mind their uninvited houseguest. Male bats carve out the cavity using their teeth, biting mostly with the incisors (front teeth) and occasionally on the edge of the canines. The skull and jaw morphology of the white-throated round-eared bat seem to reflect their cavity-creating tendencies, as this species has higher bilateral canine bite forces compared to similar-sized species. These bats will even gnaw through branches that might crisscross the cavity during building, truly customizing the perfect little termite castle. Why go through all the effort? Despite sharing the roost with a literal insect, *Lophostoma* roosting in termite mounds have significantly fewer ectoparasites like wing mites and bat flies compared to ecologically similar species roosting elsewhere. Active termite nests also have ideal thermal environments, with higher and more stable temperatures recorded in active termite nests compared to dead nests or tree cavities.

Unusual roosts can lead to unusual bodies. A handful of bats in the Vespertilionidae and Molossidae families have evolved unusually flattened skulls, with almost no forehead and very small braincases. All these flat-headed bats roost in narrow places, such as inside tight rock crevices or under tree bark. For example, the lesser bamboo bat in Malaysia roosts insides the hollow stalks of bamboo plants, entering through the openings created by leaf beetles.

## Bat Toilets and Toilet Bats

The rainforests of Borneo are home to an amazing diversity of plant life, including more than 3000 species of trees, more than 1700 species of orchids, and about 50 species of carnivorous pitcher plants. These aptly named plants have an upright, pitcher-shaped receptacle filled with liquid. Insects attracted to the edge of the pitcher face a perilously slippery surface. Those that slip in drown and are dissolved by the enzymes in the liquid at the bottom, providing important energy and nutrients to the plant.

The pitcher plant *Nepenthes hemsleyana*, however, has worked out another arrangement. The top of its pitcher has a long, concave shape that is easily detected by their mutualistic bat partner, Hardwicke's woolly bat. When the woolly bats use the pitcher plant as a day roost, the plant gets a few meals of nitrogen-rich bat poop—essentially acting as a little bat toilet. Nitrogen is an important nutrient for plant growth, and researchers found that pitcher plants that housed bats had about one-third higher nitrogen contents than those that didn't. The pitchers of *N. hemsleyana* are also perfectly bat shaped and contain lower levels of the digestive liquid normally used for dissolving insect prey. The woolly bat has also evolved structures to help it cling to the inside of the pitcher without damaging these roosts. Compared to similarly sized bats, woolly bats have proportionately larger thumb and foot pads that are more effective at sticking to the pitcher surface.

While these bats have evolved a win-win situation with plants, getting a handy roost and toilet combined, some bats have been found roosting in actual latrines. Pit latrines are used as an inexpensive way to deal with human waste in many parts of the world. A basic pit toilet consists of a large, deep vertical pit

▲ Several species of big-eared bats (genus *Lophostoma*), including this Davis's round-eared bat, have been documented roosting inside active termite mounds.

► Bats in the same genus as this Egyptian slit-faced bat have been spotted roosting inside of pit latrines in Africa.

covered over by a slab (usually made of concrete) with a smaller hole. Urine and feces enter the large pit via that smaller hole, and the whole structure is usually covered by an external shelter or outhouse. In some research and safari camps in Africa, bats have been found to consistently roost at the top of the pit walls or on the underside of the latrine slab. Using a remote-triggered camera and flash inserted into the latrine drop hole, researchers at Ruaha National Park in Tanzania surveyed the bats of the pit toilets. They found about thirty individuals across seven toilets in the camp, mostly some species of slit-faced bats (genus *Nycteris*) and one heart-nosed bat. While other published records of bats roosting in toilets are scarce, there is at least one other record of heart-nosed bats roosting in a pit latrine in Ethiopia. It's thought the pit latrines mimic the conditions of caves and hollow trees where the bats normally roost, including stable temperatures.

## Bats of a Different Color

Whether cuddled up with a conspecific or by themselves, tucked in a rock crevice or sheltered beneath a leafy canopy, bats are at their most vulnerable when roosting. Visually blending into their roosting background helps keep bats hidden from predators.

► FROM TOP TO BOTTOM: A cluster of Honduran white tent-making bats huddle in their roost tent.

A northern ghost bat roosts underneath a palm frond during the day. Its bright white coloration reflects green, helping the bat camouflage against its leafy background when viewed from below.

Face stripes like the ones on this tent-making bat might help bats camouflage by breaking up their outline when roosting in leaves.

Bats are not necessarily known for their coloration, but many species and lineages demonstrate striking facial and body markings, including spots, stripes, collars, and countershading in a range of golds, reds, yellows, and white, all of which might help hide bats in their roosts. Spots and stripes can act as disruptive coloration, breaking up the body outline and helping direct the attention of visual predators elsewhere. Bats that roost in more open habitats, like against tree bark or under leaf tents, are more likely to have evolved patterns such as stripes or neck bands. Many Neotropical tent-roosting bats are also distinguished by having yellow in the skin of their ears, noses, and wings, mimicking the yellow hues that result when sunlight filters through green leaves. In the Honduran white tent-making bat, this yellow coloration results from the integration of carotenoids—orange and yellow plant pigments—from their diet into their skin; studies suggest that the coloration may also be involved in social communication.

## Sleeping Through Winter

For insect-eating bats living in temperate regions, dropping temperatures and periods of ice and snow bring a rapid decrease in food availability. As flying animals with rapid heartbeats and high metabolisms, bats need a steady intake of energy to maintain their normal activity level. To get through winters and other periods of low food availability, such as summer cold snaps or rainy weather, many bats can put their bodies on pause until conditions improve. Torpor is a physiological state in which bats maintain their body temperature at some level below normal. Bats are not the only animals capable of this reversible depression of body temperature, also called hetero-thermy. In addition to Chiroptera, five other orders are known to have species capable of heterothermy, including shrews, rodents, marsupials, carnivores, and even small primates. The period over which bats spend in torpor varies with species and environmental conditions, and many temperate bats enter torpor as conditions demand.

Hibernation differs from torpor bouts in both duration and how much the body temperature is depressed. In hibernating bats, body temperature may drop to just above freezing, although it usually bottoms out about 1°–2°C (1.8°–3.6°F) above ambient temperature. Reducing body temperatures even

temporarily can lead to major energy savings; a 10°C (18°F) drop in body temperature results in a 50 percent reduction in metabolism. Hibernating bats still need to arouse periodically, raising their body temperatures back up to normal for short periods of time to replenish water and clear out metabolic waste. These arousals are energetically expensive, using up any fat stores saved by the bats the prior fall.

Hibernation roost climate is a critical consideration for bats, both during those periodic arousals and for ensuring they can lower their body temperatures enough to save energy in the first place. Therefore, bat hibernation locations, called hibernacula, need to meet certain conditions of temperature, humidity, and air flow while also being hidden from outside disturbances. In temperate climates across the globe, caves are popular places for bats to hibernate, though many bats will also hibernate in cavelike mines, tunnels, and culverts.

In North America, most hibernating bat species spend the winters in different locations and roosts than where they spend the summer. Starting in late summer and into the fall, bats begin moving to a communal hibernaculum. This period is associated with mating behavior at and outside the hibernation roosts—called swarming—and hyperphagy, where bats try to eat as much as they can to store fat before the winter. Bats also move around in the winter roost as temperatures and humidity change throughout the season, clustering with other individuals to help maintain the stable temperatures most optimal for hibernation.

Caves might be a popular hibernation roost, but they are far from the only place that bats will roost for hibernation. Following a series of anecdotal observations of Ussurian tube-nosed bats under or near the surface of snow during

► Cave myotis cluster tightly while hibernating in a north Texas cave.

the winter in Japan, researchers set out on an unusual quest to find bats in the snow. After more than 300 hours of searching, the researchers observed 36 live bats roosting in either small depressions or cylindrical dents in the surface of the snow. Based on the physics of ambient temperatures and snow melting rates, they concluded that the bats found in the snow had most likely spent the winter there, becoming visible only when temperatures began to warm enough that the snow melted away. It's unclear how common this snow-hibernation behavior might be in this species, as the bats will also hibernate in tree cavity roosts.

It might seem strange at first, but hibernating in the snow may offer bats a few advantages. One is the conservation of water. Even with reduced metabolisms, bats lose a lot of water during hibernation to condensation, and water recovery is one of the reasons that they need to periodically wake up during the winter. Hibernating in the snow might minimize the cost of replacing that lost water. Most bats are not able to withstand freezing temperatures, even when in torpor. It remains to be fully tested, but the Ussurian tube-nosed bats might be among the few that are able to survive subfreezing temperatures in a super-cooled state.

Eastern red bats also don't use caves or cavelike hibernacula during the winter. Instead, they roost in tree foliage during relatively warm winter days and move to the ground to huddle in leaf litter when temperatures drop and they need more insulation. This behavior occurs more frequently in more southern areas of North America, where winters are generally mild and there are occasional warming periods during which bats might be able to wake up and hunt. Eastern red bats are also an excellent example of a bat that employs a more hybrid approach to surviving the winter: while the bats can and do spend some time hibernating, they will also undertake long-distance seasonal movements more consistent with migration.

# Long-Distance Travel

Instead of hibernating, some temperate bats migrate. For example, come autumn the hoary bat and silver-haired bat begin to leave their summer breeding grounds, heading south for milder weather conditions. These bats are adapted to roost in trees, which are not as well insulated against harsh weather as caves. Hibernating bats in some regions will travel long distances from their summer grounds to their winter roost.

Just how far will bats go? For many bats, we still don't know exactly. Most bat-tracking research relies on mark-recapture methods, such as banding or PIT tags. These methods require that the band or tag be recovered in a new location, which can be challenging when dealing with small, elusive creatures like bats. Large-scale banding efforts in the early twentieth century form the basis for much of what we know of bat movements in North America. From banding data, we know that Mexican free-tailed bats will move from summer grounds in the southeastern and south-central United States to coastal Mexico, traveling as far as 1800 km. The tiny Nathusius's pipistrelle, a bat that only weighs about 8 g, has been recorded making transcontinental movements from Russia and Latvia to the French Alps and Spain, distances exceeding 2300 km. Some bat populations seem to undergo partial migration, where some individuals stay put or move shorter distances than others. In the common noctule bat, a European migratory bat species, some individuals breed and hibernate in the same location, some travel over 800 km, and others do something in between.

Advances in technology, like GPS units that are small enough to be carried by bats, are helping uncover details of where bats are moving during migration. In the redwood forests of northern California, hoary bats can be reliably captured in the fall as they migrate through the area. Based on previous work, it's generally assumed that these bats are on their way south, taking advantage of coastal forests as commuting grounds. After recapturing a few individuals during this annual flurry of hoary bat activity, Ted Weller of the US Forest Service and colleagues wanted to test if that was the case or if maybe some of these bats were simply undergoing partial migrations. I had the good fortune to be a part of this project as we captured and attached miniature GPS units to eight bats in the fall of 2014. Because the tags had to be miniscule to fit on a bat, we needed to recapture the bats to track their movements. One recaptured bat had moved more than 1000 km in about a month, going up to southern Oregon and over to Nevada before making his way back toward the site of capture. Another bat flew at least 68 km in a single night, having left the capture area before returning. The same study also found evidence that hoary bats are capable of hibernating, with one individual showing a period of inactivity lasting about forty nights based on body temperature data.

◄ A hoary bat is released after being fitted with a GPS tag.

▶ In addition to their silver-tipped fur, hoary bats also have striking coloration on their wings.

# Surfing the Green Wave

Instead of moving to hibernacula or better weather, some bats migrate to track seasonal changes in food availability. This is the case for the North American nectar-feeding bats: the Mexican long-nosed bat, the lesser long-nosed bat, and the Mexican long-tongued bat. These bats move long distances (as far as 1200 km) from their winter habitat in south-central Mexico to northern Mexico and the southwestern United States. During their journey, they follow a so-called nectar corridor of cacti and agave, timing their movements northward with the blooming of these flowers.

In other migrating mammals, like caribou and wildebeest, long-distance movements are also timed to coincide with changes in plant growth and nutrients, and their movements are described as "surfing the green wave." Turns out, fruit-eating bats in Africa also track food availability in a similar way, with bat seasonal movements matching environmental proxies for plant growth and (assumed) fruit availability. For the straw-colored fruit bat this movement culminates in the world's largest mammalian migration, as an estimated 10 million bats make their way to Kasanka National Park in northern Zambia. Occurring between late October and mid-December every year, fruit bats literally fill the skies each night as they go in search of fruits like waterberries, mangos, and loquats that are in high abundance.

Straw-colored fruit bats aren't the only bats that move long distances as they seek fruiting trees. Flying foxes in Australia also move long distances among a network of roosting trees in search of food. Unlike migration, which consists of seasonal, unidirectional movement, flying foxes are more nomadic. In a study that tracked the movements of little red flying foxes, grey-headed flying foxes, and black flying foxes for 5 years, researchers observed bats moving between more than 750 different roosts. Little red flying foxes were tracked moving as far as 6000 km per year, zigzagging over the forests as they searched for flowering and fruiting trees. These wandering movements exceed the longest distances travelled by migrators like caribou and wildebeest, who only travel between 1200 and 2900 km each year.

▼ This Mexican long-nosed bat was captured during summer roost surveys at the only known maternity colony in the United States.

▲ Straw-colored fruit bats are widespread throughout Africa and undertake seasonal migrations that follow the fruiting of favored trees.

◄ Little red flying foxes in Australia don't undertake seasonal migrations, but they do travel long distances over the year as they seek flowering and fruiting trees.

♥

# LOVE

## Chapter 9

Birds do it. Bees do it. And bats do it. Let's talk about sex and bats.

Bat mating behaviors are like those of many other mammal species. Bats reproduce sexually, meaning that sperm from a male and an egg from a female must combine to form an embryo. Categorizing bat mating systems and patterns is difficult, however, due to the huge diversity, variation within species, and large gaps in knowledge for many species worldwide. Some bats are polygynous, with males mating with multiple females whom they guard from other males, whereas other bat species live more monogamous lifestyles, with lots of variation in between.

In many species of bats, males have specific features that distinguish them from females, resulting in sexual dimorphism. These differences can be reflected in appearance, such as different coloration or fur patterns, or in behavior, like songs or courtship dances. From mixing perfumes to singing unique love songs, bats have more than a few tricks up their wings to attract the perfect mate.

► Wrinkle-faced bats have courtship rituals as strange as their faces.

# Eau de Chiroptera

I have a weird confession to make: I love how bats smell. Some bats give off a smokey, spicy aroma, reminiscent of sugar-crusted bacon. Others have a faintly floral scent, like a slightly rotten bouquet of flowers. Still others emit the unusual combination of animal musk and mint. Other bat scents are almost impossible to describe, but all do have one thing in common—they are produced almost exclusively by the males of a given species. Of the twenty-one families of bats, at least fifteen have species that demonstrate sexual differences in odors or odor-producing glands or structures.

## Are you gland to see me?

In many bat species, these scents originate from specialized glands on males. Gland locations include the throat, chest (also called gular glands), shoulders, armpits, back, wings, and around the genitalia. Males of many free-tailed bat species have sebaceous gular glands, which are most prominent during breeding periods and may help communicate information about a male. In a study of the Florida bonneted bat, males with open gular glands were heavier and had bigger testes than males with closed gular glands, even when standardized by forearm lengths. Some species, like the Mexican free-tailed bat, have been observed rubbing the secretions of these glands on roost walls, possibly

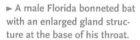

► A male Florida bonneted bat with an enlarged gland structure at the base of his throat.

to mark territories during mating season. Meanwhile, the Sinaloan mastiff bat might just mark female bats directly, leaving an oily, stinky spot on their backs (gross). It's not clear if this is simply an accidental transfer of these secretions or if males are actively marking females as a signal to other individuals of the same species.

Male greater spear-nosed bats also develop large chest glands during the mating season. These glands produce a thick white secretion with a pungent odor. Greater spear-nosed bats generally form harems of ten to twenty-five unrelated females who are all guarded by a single male. Although this male gets first access to mating with these females, he must also defend his females from neighboring harem males and roaming bachelor males. Gaining access to and defending these female groups can be difficult, with intense fights leading to wounds on the males' faces, bodies, and wings. When comparing the chemical composition of gland secretions between bats, researchers found that harem males and bachelors males produced significantly different combinations of chemicals and that chemical signatures also differed among harem males and among bachelors. These findings suggest that these odors might serve to communicate information about mating status and individual identity. Females might also be using chemical signals to choose between males for mating. Alternatively, males might use this information to evaluate who and when they want to fight.

▲ Greater spear-nosed bats roost in an abandoned building near Gamboa, Panama.

## Hair-raising scents

In some bats, glands are surrounded by special types of hairs called osmetrichia. These so-called scent hairs are structurally different from other body hairs, with grooves or large chambers between the hair cuticle scales. These grooves and chambers either hold or move secretions up and away from the gland itself, helping further disperse the odors from the bat's body. A dramatic example is the rock-and-roll mohawk (technically called the interaural crest) of male Chapin's free-tailed bats. These long white hairs form a spiky crown on the bat's head and are thought to help disperse odors from a glandular structure on the head.

Male yellow-shouldered bats have conspicuous epaulettes, patches of dark brown and orange on their shoulders. Darker and longer than the surrounding fur, these shoulder patches are also associated with scent glands that give the fur a waxlike coating. The odors from these shoulder patches can be very strong. Chemical analysis of these shoulder secretions in the northern yellow-shouldered bat revealed odor profiles dominated by terpenes and phenolics. One of these compounds is linalool, which has a scent described as "floral and spicy" and is commonly produced by plants like lavender, bergamot, and citrus. While adult males have unique scents compared to juvenile males and females, it's still unknown exactly how yellow-shouldered bats use these scents in a mating context.

Male bats of several species in the Pteropodidae also have unusual fur patterning, such as neck ruffs and shoulders with extra pizazz. Epauletted fruit bats have skin pockets on their shoulders, usually associated with long, bright white hairs. Although these hairs resemble osmetrichia, no secretory structures or glands have been found in these skin pockets. Of course, this doesn't rule out an odor-signaling role of these epaulettes, but scientists think the hairs might be more of a visual mating cue in these non-echolocating fruit bats.

## Master mixologists

Microbes play a part in refining or modifying the actual odors associated with an individual or a given species. Researchers from the University of Los Andes in Venezuela used gene sequencing to identify the bacteria associated with the epaulettes of yellow-shouldered bats, and they found that males of two different species had different types and combinations of bacteria on their shoulder patches.

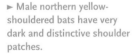

▶ Male northern yellow-shouldered bats have very dark and distinctive shoulder patches.

Males can add even more customization to their scent signals by mixing secretions from various parts of the body. During their two-month breeding period, male long-nosed bats in Venezuela develop conspicuous, greasy patches between their shoulder blades. These patches are accompanied by a strong, musky odor but there is no nearby scent gland. Instead, males build up these dorsal patches with a complex smearing behavior, using their feet to gather secretions from their mouth, head, penis, and anus before depositing them onto their back. Chemical analysis of these dorsal patches identified some compounds that are natural insecticides, which led researchers to hypothesize that these glands might both help males fight off pesky parasites like bat flies and advertise their parasite-free status to potential mates. In fact, males with dorsal patches did appear to have fewer parasites than males without dorsal patches. Furthermore, females showed strong preferences for males with dorsal patches, lending support to this hypothesis.

Long-nosed bats aren't the only ones that spruce up their appearance and scent each breeding season. Researchers studying fringe-lipped bats in Panama had observed an unusual orange substance forming a hard, flaky crust on the forearms of males. This mysterious crust was also characterized by a sharp, pungent odor and had not been previously described in the scientific literature. Further research by Victoria Flores (at the time a predoctoral research fellow at the Smithsonian Tropical Research Institute) revealed that males with this orange crust had enlarged chest glands and testes. Males with forearm crust were also in better body condition than crustless males, and males with the biggest area of crust also had the highest levels of testosterone. However, there was still the major question of where this crust came from, as no gland is present on the outer edge of the forearm where the orange crust is found.

Using video cameras supplemented with infrared lighting, researchers spied on fringe-lipped bats both in their natural roosts and in captivity. It turns out that males prepare this crust with a very specific sequence of behaviors, scratching at their body and chest glands with their foot, licking the foot and then repeatedly licking their forearm. Though the mystery of where the orange crust came from was solved, it was still not clear why males spend time creating this crust. The initial hypotheses centered around female choice. Do males form this crust to attract females, and are females making choices based on the size or smell of the crust? Females did not show strong attraction to the odor associated with the crust, but males with forearm crust did show a preference for the scents of crustless-males. These findings indicate that, instead of attracting females, the forearms crust serves as a signal to other males, allowing males to evaluate their competition.

The true chiropteran kings of perfume mixing are the greater sac-winged bats. True to their common name, male sac-winged bats have a pocket of skin situated just above their elbow on either side. This sac is bordered by two muscle ligaments, which allow the wing sac to be opened and closed. The wing sacs frequently contain a yellowish, strong-smelling liquid, despite the lack of any secretory or glandular tissue in the wing sac itself. That's because each afternoon males spend about an hour cleaning out and then refilling these wing sacs, blending a combination of genital and gular gland sections together with their mouths. The wing sac odors consist mainly of compounds with aromatic carbon rings (such as indole, which is also found in jasmine oil and coal tar) and fatty acids. So why do males go through all this trouble every day?

Sac-winged bats roost vertically among buttresses of large rainforest trees in Central America, as well as in well-lit hollow tree cavities and in human-made structures like sheds and storehouses. This has made them relatively easy to study, with some research programs tracking and observing the behaviors of colonies for years, even decades. Colonies consist of small subgroups made up of a single territory-holding male and several females. Territory-holding males are more likely to mate and reproduce with associated females than those without a territory (so-called satellite males). During the mating season, territory-holding males perform courtship displays, hovering in flight and wafting the odors from their wing sacs toward females. These courtship displays—combined with

▶ A male fringe-lipped bat emerges from a roost in Lamanai, Belize. Note the crusty, orange material on its left forearm.

◄ A male greater sac-winged bat leaves its roost in the evening. Note the small sac-shaped structures visible at the elbow of each wing.

the dominance of females over males during social interactions—suggests that females are the ones doing the choosing and odors play a role in that decision.

What can wing sac odors tell the females about a bat? Potentially a lot. Sac odors change as males reach sexual maturity, with several chemical substances found only in the wing sac secretions of reproductive adults. Different individuals also have unique combinations of odors, so females could potentially distinguish between individuals from scent alone. Unique odors may also help prevent hybridization between similar species. The composition of wing sac odors was significantly different between greater sac-winged bats and the closely related lesser sac-winged bat, and female greater sac-winged bats showed strong preferences for the odors of their own species. Finally, scent may help prevent inbreeding. In most mammalian species, the males are the ones who leave home after maturity, whereas daughters often stay in the same area where they were born. Greater sac-winged bats are an exception to this pattern, with females leaving their home roost to join colonies elsewhere. Females show a preference for scents of males from distant populations compared to the scents of local males, which can help females avoid mating with relatives after they leave their home roost.

# Sing Me a Song

Male sac-winged bats don't just bring sexy scents to the table when trying to woo females. In addition to hover flights that waft the smells toward females, sac-winged bats are also prolific singers. Of course, singing is also important in human culture, but singing in other mammals is rare and not as well understood as birdsong. Songs and singing behavior have been documented in at least twenty bat species, spread across five families of bats (all of which also echolocate). Although some parts of bat songs dip low enough in frequency to be detected by human ears, much of the action happens out of our hearing range. Add in the secretive and cryptic lifestyles of many bats, and I wouldn't be surprised if more bat singers are uncovered in the future.

How is bat song different from other vocalizations? In the world of animal behavior, songs are generally multipart sequences of calls that are often produced in bouts. The length and complexity of a song varies, such as how many times a certain part of the song might be repeated or how different parts of the song are arranged in sequence. The individual parts of a song are referred to as syllables, and the arrangement of syllables into certain patterns is referred to as syntax.

Greater sac-winged bats sing territorial and courtship songs. Territorial songs are short songs (about 2 seconds) that can sound a bit like a high-pitched whinny to human listeners. Males break out territorial songs in response to hearing another bat approaching or leaving the roost. In contrast, courtship songs are only directed toward females in the roost and are composed of a variety of different syllables that can last up to 40 seconds. The most common element of courtship songs is the trill, a syllable consisting of a vibrato, rippling pattern. An individual male will sing a variety of songs, putting together trills, squawks, and tonal calls into different combinations, and males differ in their songwriting styles. Songwriting skills matter, as males with more elaborate and varied repertoires attracted more females in their territory, increasing their opportunity for mating.

Mexican free-tailed bats combine the territoriality and courtship aspects all into one song. These bats mate in the early spring, before or during their northward migration. Males stake out day roosts that also serve as karaoke platforms, where they sing spontaneously and often. As in sac-winged bats, these songs are highly variable within and between individual males. Interestingly, free-tailed bats are not just randomly combining syllables but are following syntactical rules in their songs. These syntactical rules give a structure to the bat songs, in the same way that we combine certain words in English to form proper sentences. Free-tailed bat songs are composed of several syllables and follow an organized pattern of three phases: the first phase consists of an introductory run of chirps; the second phase contains some combination of trills, chirps, and buzzes; and most males choose to finish their song with a combination of trills and buzzes. In observations of almost 300 Mexican free-tailed bat songs, researchers found that males never followed a buzz syllable with a trill, with common sequences including chirp-trill-chirp, chirp-trill-buzz, and trill-chirp-buzz. These combined syllables are thought to both encourage females to join them in the roost and act as a threat to ward off potential male competition.

Sac-winged bats and free-tailed bats perform their courtship songs from within the roost, but some bats broadcast from the open skies. Several species of European pipistrelles—small, insect-eating bats in the family Vespertilionidae—perform territorial or courtship songs during flight. Male common pipistrelles are particularly good at putting themselves in the spotlight and making sure females notice them. In the fall, males arrange their territories in such a way that females must pass through them on their way to winter roosts. Singing and flying is no easy task; imagine trying to sing at the same time as you sprint a 100-m dash. Breathing and vocalizing are closely linked to wingbeat patterns in a flying bat, which puts a considerable energetic constraint on flight song. For that reason, pipistrelle flight songs tend to be short, rapid bursts either contained within a single wingbeat or sung over a span of several wingbeats.

▶ This Mexican free-tailed bat song starts with a series of chirps, followed by one trill and an extra chirp before ending in a buzz syllable. Except for the last buzz, most of the song is above human hearing range.

◀ Mexican free-tailed bats are one example of a prolific bat singer.

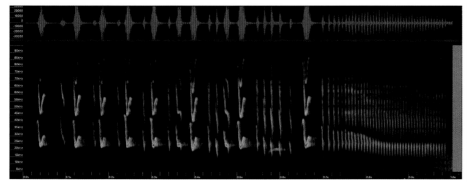

# Lek's get it on

Most patterns of bat mating and courtship display fall into two main categories: resource defense (such as territorial free-tailed bats fending off male interlopers from their day roosts, where they hope to mate with females) and female defense (sac-winged bats courting nearby females while trying to keep other males from stealing mating opportunities). Lek breeding is one of the most unique mating systems, seen in only a few species of bats. In this mating system, males come together at a central location and produce displays for the purpose of attracting females, who come visit these arenas for the sole purpose of checking out the males. Female selection in these mating systems is often highly skewed, with only a small proportion of the males achieving mating success.

A classic lek-breeding bat is the hammer-headed bat. This species is among the largest bats in continental Africa, where they feed on fruits and roost in the forest canopies of West and Central Africa. Males and females look dramatically different, with males weighing nearly twice as much as females and having a bulbous and elongated nose that earn them the hammer-head moniker. During the breeding season, males arrange themselves along riverbanks and produce loud, low-frequency honks to attract females. To accomplish these honking displays, male hammer-headed bats also have a larynx that is nearly three times larger than that of females, filling about half of their entire body cavity.

Another bat whose mating system falls into the category of lek breeding is the New Zealand lesser short-tailed bat, the peka-peka. Male peka-peka are the true bards of the bat world. During the breeding season, males gather in small tree cavities used explicitly for courtship displays, called singing roosts. Some of these roosts are occupied by only one male, whereas in others several males share the stage, with each male getting a performance slot as females visit throughout the night. If the females like what they hear, then mating might occur. Like the elaborate songs of sac-winged bats and free-tailed bats, peka-peka use four main notes, which they combine into as many as fifty-one distinct syllables. During their average 6-hour-long performances, male peka-pekas rarely fall silent, singing up to 100,000 syllables a night. This makes them true champions of song, with one of the highest sustained song outputs for either bats or birds. Furthermore, these syllables can provide listening females with information about the singing bat's size, as larger males tend to produce shorter trill downsweeps than those of smaller males.

The wrinkle-faced bat is another species with a unique courtship behavior that's consistent with lek breeding. Both males and females of this iconic bat species have some of the strangest facial morphology of all the New World leaf-nosed bats, full of folds and wrinkles. Males also have a fold of skin under their chin that can be raised up in the style of a face mask, which was assumed to be related to mating behaviors. Researchers in Costa Rica discovered an aggregation of males at tree perches, where they were hanging with their skin masks raised up over their faces. At the same time, the males fluttered their wing-tips and emitted spontaneous, high-frequency social calls. When another bat would approach, males began flapping their wings vigorously and singing loud, low-frequency whistling songs, including a trilled syllable often heard in other bat songs. These observations offer a tiny glimpse into the sex lives of wrinkle-faced bats, though there is still much to be uncovered in this species, including if this is a widespread mating behavior and what parts of the court-ship display females might find the most interesting.

▼ The latticed wing pattern of the wrinkle-faced bat might be part of the wing-and-calling display males use to attract females.

▲ A close-up view of the male's wing pattern

## Food for Sex

Many of the courtship behaviors seen in bats—from sexy scents to complex songs—are likely aimed at females to signal the quality of a male as a poten-tial mate. These honest signals can facilitate the evolution of certain traits via female choice, as females might prefer certain traits in males, which then get passed down to their offspring and increase the reproductive success of the next generation, continuing the pattern. These signals can also be a way for males to avoid potentially costly fights by communicating about their body size or condition to other males.

Another male strategy is to demonstrate how good they are at acquiring resources, such as food. In Egyptian fruit bats, this strategy is reflected in males trading food for sex. In a captive colony of bats at Tel Aviv University, researchers observed that bats show two main strategies for acquiring food: going out and finding food themselves (producers) or stealing food from the mouth of another bat (scroungers). Males tended to be the producers in the colony, with females frequently nabbing food right out of a male's mouth. While these interactions could be quite aggressive, researchers noted some patterns among who females stole from and how often. They later analyzed the paternity of offspring from these seemingly thieving females and found that females gave birth to the young of males from which they had taken food. These observations may be evidence of sexual selection in which females receive resources in the form of nutrition and males gain access to mating. So far, these behavioral interactions have only been observed in captivity, so it is unclear if these patterns also apply to Egyptian fruit bats mating in the wild.

## Mating Season and Swarming

For many hibernating bats living in temperate North America and Europe, mating occurs primarily during autumn swarms. During the fall, many hibernating bats begin to leave their summer grounds for hibernacula, where they will curl up for the winter. But the bats don't just head straight for the winter nap. Instead, they spend several weeks in and around the hibernacula, fattening up on insects, inspecting the winter roost, and socializing. Nighttime just outside the hibernacula often consists of large numbers of bats circling, chasing, vocalizing, and mating. This swarming behavior is important for maintaining gene flow among different populations of a bat species across its range. Bats that spend summers in several different places might all converge on the same hibernation location, resulting in the mixing of genes adapted to different areas in the next generation of offspring.

In migrating temperate bats, fall is also thought to be mating time. Due to the particularly elusive nature of many migrating bat species, observing their mating in action is almost impossible. Instead, researchers estimate

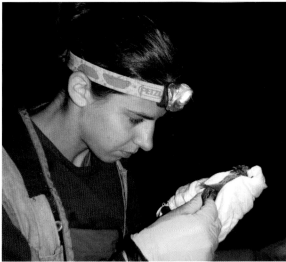

reproductive readiness by examining the testes of male bats. When reproductively ready to mate, male bats display enlarged testes and distended epididymides (the tubes by which sperm moves from the testes to the penis). Although it seems that these migratory corridors are important sites for mating in species like hoary bats and red bats, there are still a lot of unknowns.

In coastal northern California, a migratory hotspot for hoary bats, the sex of bats captured skews overwhelmingly male. During my time working in this area, this led to many nights out in the field with my colleagues and I taking bets on whether we would catch any females and how many we might capture. Our inability to capture females does not necessarily mean they weren't there or that males weren't just on their way to find them before we waylaid them with our mist nets. But it does highlight how much there still is to uncover about bat mating behaviors.

In tropical and subtropical regions, bat mating patterns are even more of a mystery. Instead of experiencing winter and summer temperature changes, tropical regions around the world see annual variations in rainfall. Based on captures of pregnant or lactating females and reproductively ready males, mating behaviors of tropical bats seem to fall into a few general categories. Some species—like the long-winged tomb bat, the short-nosed fruit bat, and the common vampire bat—may breed year-round. Other species likely breed only once or twice a year, with mating often tied to the availability of foods such as flowering or fruiting trees.

▾ The Indiana bat is a species that likely mates in the fall prior to hibernation.

▴ The author checks testes and epididymides of a male hoary bat during fall migration surveys in northern California.

# Kinky bats

Competition between males doesn't necessarily end after mating. Across much of the bat world, females don't just mate with one male. Even in mating systems where one male defends his group of females from other males, paternity analysis of pups indicate that females are also seeking mating opportunities elsewhere. What's a male to do? Multiple mating, combined with the potential for sperm storage and delayed fertilization in many temperate bats, sets the stage for what's referred to as postcopulatory sexual selection. One potential adaptation related to this postmating competition is species or seasonal variation in the growth of penis spines.

In some bats and a variety of other mammals, the rounded tip of the penis (the glans penis) is covered in keratinized spines of varying sizes and shapes. Migratory tree bats like hoary bats and eastern red bats have particularly long penis spines that are visible to the naked eye—up to 1 cm in hoary bats. That might not seem particularly long, but when compared to the males' overall size, it comes out to about 6 percent of their entire body length. Penis spines are also found in some species of free-tailed bats and tube-nosed bats (genus *Nyctimene*). Variation in the length, distribution, and shape of penis spines can even help researchers discriminate between closely related species, such as very similar species of dog-faced bats (*Cynomops*).

Scientists think that penis spines are linked to sexual selection, but exactly why some bats have them and others don't is still a bit of a mystery. In bats that mate with multiple partners, penis spines might allow a copulating male to scoop out the sperm from a previous mating encounter. Another hypothesis suggests that the hoary bat's extra-long penis spines help prolong mating by acting as a sort of locking mechanism, keeping mating bats together as they attempt to mate in flight. Unfortunately, further observation of both male and female anatomy and mating habits are needed to really understand the relationship between penis spines and sexual selection in bats.

Giant penis spines aside, male bats may have other strategies for increasing their likelihood of fathering offspring with a given female. Male Indian flying foxes have been observed performing oral sex on females, licking at the vagina prior to mating. The duration of male licking and mating were positively correlated, with longer premating sessions of cunnilingus resulting in longer

◄ Black flying foxes roost in trees, where their large size and gregarious nature make it easier to observe their behaviors, including those related to mating.

mating bouts. Black flying foxes have also been observed performing cunnilingus prior to mating, though the resulting time spent mating was not quantified in those observations.

Female licking of the male's genitalia during sex can also increase mating time, as observed in short-nosed fruit bats. Researchers in China found that females regularly licked their mate's penis during sex, with each second of licking resulting in about an extra 6 seconds of mating. The researchers hypothesized that this behavior could be a way for females to assess males based on chemical cues and increase the likelihood of fertilization by the chosen males. Further observations and study would be needed to truly disentangle these hypotheses. It is also worth noting that the researchers for this study won the 2010 Ig Nobel Prize for Biology, a satiric prize meant to celebrate silly or seemingly trivial research discoveries.

# GROW

## Chapter 10

Raising a baby bat is a lot of work. Bats give birth to live young, called pups, that are born mostly hairless, flightless, and often blind. Despite their generally small size and visual similarities to quickly reproducing animals like mice, bats live life in the slow lane. Most bats only give birth to one pup at a time, though there are some bat species where twins, triplets, and even quadruplets are not uncommon. As with many mammals, bat mothers take on most of the work when it comes to bearing and raising young. Of course, not all bat families are the same, and bats species vary in everything from how long baby bats stay in the womb to how they learn to fly and forage.

▶ An epauletted fruit bat roosts with her pup.

# Pregnancy and Birth

In humans, after sperm fertilizes an egg, the new zygote gets immediately implanted into the uterine wall and begins development. In many other mammals, however, reproductive delays occur at different stages of this process. One such pause is the delayed fertilization seen in many species of hibernating bats. Following mating in the fall, females store sperm in their bodies over the winter, waiting to initiate fertilization and pregnancy until the spring. Another type of pause is a pre-implantation delay, seen in bent-winged bats. In this species, fertilization happens shortly after mating but instead of immediately becoming implanted in the uterus for further development, the newly fertilized bundle of cells (called a blastocyst) remains in a suspended state during hibernation. This bundle of cells doesn't implant in the uterus until after the mother has woken up in the spring.

In some species, these delays always happen as part of the bats' normal reproductive cycle. They may have evolved as ways to maximize reproductive success, helping time pup growth and development with periods of better food supply or thermoregulatory conditions. This hypothesis is consistent with the observation that reproductive delays are most often observed in temperate species that encounter significant seasonal variation that can impact survival. In other species, these delays are not a standard part of the reproductive process but might instead be mediated by environmental conditions. Exactly what environmental factors are responsible for this variation is unclear and likely varies among species. One hypothesis about reproductive delays is related to hibernation and torpor. During torpor, all metabolic activity in a bat's body is slowed, including fetal growth. This idea is supported by observations that bat gestation times increase during cold weather and long periods of torpor in pregnant bats.

For their size, bats have unusually long gestation times that span from 2 to 7 months in different species. The small, insect-eating vespertilionid bats tend to have the shortest pregnancies, ranging between 50 and 80 days. Other families of primarily insect-eating species trend closer to 90 days, such as free-tailed bats, bent-winged bats, and horseshoe bats. Overall, species living in the tropics tend to have longer gestation periods than those of temperate species. Fruit- and nectar-feeding species in both the Old World and New World tropics have some of the longest gestation periods, ranging from an average of 4 months in

phyllostomids to closer to 6 months in some flying fox species. Interestingly, the bat species with the longest pregnancy is not a flying fox or even a fruit bat. It's the common vampire bat, who gives birth to a single pup after a 7-month pregnancy.

Why do bats have such long pregnancies? Relative to other mammals of similar sizes, fetal development is slow in bats. Bat babies are also big babies, at least relative to the size of mom. Pups of the little Japanese horseshoe bat can weigh as much as 45 percent of the mother's weight at birth. On average, newborn bat pups weigh about 20 percent of their mother's total body weight. That would be like an average woman giving birth to a 30-pound baby! Interestingly, bat species with larger bodies like flying foxes tend to give birth to relatively smaller pups (only about 10 percent of the mother's weight), whereas small bats give birth to the relatively largest pups.

For most bat species, females give birth to only one pup at a time. Considering the wide variety in other traits, such as body size, diet, and habitat, it's noteworthy how consistent this life history trait is across species. Some species of vespertilionids, such as the evening bat and pallid bat, consistently give birth

▲ A visibly pregnant fringe-lipped bat leaves her roost for the evening.

to more than one pup; usually this means twins. Red bats have some of the largest litters, giving birth to as many as five pups at a time, while averaging three in a litter. Not surprisingly, in bats that do give birth to more than one pup at time, the size of each pup is smaller relative to the mother as compared to single-pup species.

If you're wondering exactly how bats can give birth to such relatively large babies, part of the answer is in the pelvis. In general, the bat pelvis is like that of a human, a set of two curved, wing-shaped hip bones (os coxae) that are connected in the back by the sacrum at the base of the spine and in the front via the pubic symphysis. The shape of the pelvis is sexually dimorphic, showing significant differences between males and females. Female bats have a gap between the two hip bones where they would normally meet in the front. This gap is spanned by a cartilaginous ligament, which may facilitate the ability for the pelvis to stretch during birth and accommodate larger newborns.

Hanging upside down and large babies can pose some challenges for bat moms when it comes time to give birth. In some bats, the mother remains upside down while giving birth, catching the newborn pup with her wings and tail membrane. In other bat species, the mother repositions herself head upward, with all four limbs (feet and thumbs) on the roost surface to give birth. Some large fruit bats hang feet-downward, holding on to a branch with their thumbs.

Bat pups are usually born breech, with the feet being the first to emerge, though head-first births have been observed in some horseshoe and short-nosed fruit bats. For breech presentations, as the pup is pushed out it uses its feet to grasp onto the mother's stomach, assisting in its own birth. Females lick and bite as the pup is born to remove the amniotic sac and sever the umbilical cord. The placenta is delivered shortly after the pup and is usually consumed by the female.

The time in labor depends on species, but seems to take between 30 minutes and 3 hours. While pregnant bats are often found roosting together, giving birth is a solitary process and females may even move away from the group when in active labor. In one extraordinary example, female Rodrigues fruit bats may help each other during birth. An attending female bat was observed grooming a female in active labor, helping move the laboring female into an upright position, and grooming the emerging pup.

# The Mom Hangout Spot

In many bat species, especially those in temperate areas, females congregate in groups to give birth and raise their young. These maternity colonies can provide important benefits to both expecting and nursing bat mothers as well as for their newborn pups. One important benefit of maternity colonies is thermoregulation. Pregnancy and lactation are energetically demanding in mammals—maybe especially so in bats. While many species of bats can handle temporary drops in temperature or food supply by entering energy-saving torpor, pregnant and nursing females tend to enter torpor less than males do. It's thought that the short-term benefits of saving energy during torpor are outweighed by the costs of longer pregnancy and slowed pup growth, especially in areas with short growing seasons. Clustering together during periods of cooler weather can help bats maintain higher body temperatures while expending less energy. While less common and not as well studied, maternity colonies in tropical foliage-roosting bats may also be driven by group thermoregulation. For example, Peter's tent-making bats form loosely structured female groups during pregnancy and birth, which may reduce energetic costs during this metabolically expensive time in a female bat's life.

The thermoregulatory benefits also extend to the mostly naked newborn pups, who are unable to thermoregulate on their own in the first few days following birth. In species that leave their young in the roost while they forage, pups often clump together into creches to maintain their body temperatures

▶ A maternity colony of Yuma myotis gathers in the attic of a historic hotel in northern California.

▲ A southern bent-winged bat mother with a large group of pups in a maternity colony.

► FROM TOP TO BOTTOM:
The nipple of this great
fruit-eating bat is enlarged and
balding, indicating that she is
currently nursing a young pup.

This common vampire bat pup
has a cute, inquisitive face.

A mother vampire bat nurses
her pup.

while mom is out hunting. Bat mothers also take an active approach in ensuring appropriate microclimates for developing babies by choosing roosts with the right temperature and humidity. This may be particularly important for bats that roost alone or in smaller groups, such as species that roost in tree cavities or crevices.

Bats that congregate in larger roosts such as caves or attics can take advantage of natural microclimatic variation throughout the roost. Depending on the shape of a cave, the underlying geology, and patterns of airflow, different cave areas have different temperatures. Females may choose where to leave their pups based on this natural variation. In some cases, the presence of bats can actually influence the ambient temperature. Nursery areas in breeding caves of the common bent-winged bat in Australia were warmer than expected based on seasonal and airflow patterns. Bats roosting in human-made structures like bat boxes and attics have similar preferences. Attic maternity roosts of big brown bats in Pennsylvania could reach daytime temperatures as high as $55\,°C$ ($131\,°F$), and occupied attics were consistently warmer than unoccupied ones. Occupied attics also had wide temperature gradients, being cooler in some spots and warmer in others. By moving between these different microclimates, bats can regulate their temperatures and avoid heat or cold stress.

# Bat Babies

Maternal investment in the young extends significantly beyond pregnancy in bats, and bat moms devote a substantial amount of time to caring for the pups after birth. Even though bat pups are relatively large, they still require a lot of parental care. Compared to the growth rates during pregnancy, pups grow and develop relatively quickly in most bat species. Especially rapid growth, early weaning, and shorter time to flight are seen mostly in temperate species, likely reflecting the importance of achieving maximum growth and fat reserves before the weather changes and it's time to either hibernate or migrate.

Following birth, bat pups rely on milk from their mother to survive. In most bat species, females have one set of nipples, though bats that give birth to multiple pups (such as red bats) have an extra pair of nipples to support newborn pups. During most of the year, bat nipples are small and difficult to find. During lactation, nipples become larger and have less hair around them because of nursing. For scientists studying bats, looking at the size and state of a female's nipples is important for determining reproductive status.

Nursing frequency and duration depend on the bat species. In some, new-born pups remain attached to their mothers almost constantly, including during flight. In other species, females nurse on and off throughout the day and night. The nutritional composition of the milk is an important factor that can influence how often moms need to nurse their pups. Bats that produce milk with higher concentrations of energy-rich nutrients like fats and protein nurse less frequently than those with less nutritious milk. Milk composition in bats depends a lot on their diet: insect-eating bats produce milk with higher fat and protein content than that of fruit- and nectar-feeding species.

Female bats nurse their young somewhere between 21 and 90 days. Weaning usually coincides with when juvenile bats begin flying and hunting on their own. One major exception, however, is the common vampire bat. After the unusually long 7-month pregnancy, vampire bats nurse and care for their young for another 7 to 10 months. This extended pup care in vampire bats also includes introducing blood into the pup's diet via regurgitation, a behavior not observed in any other group of bats.

# Oh Mother, Where Art Thou?

For bats that roost alone or in small groups, finding their own young for nursing is an easy task. In Bracken Cave, home to 20 million Mexican free-tailed bats and the largest known bat colony in the world, it's a little more challenging. Here, creches can be as dense as 4000 pups per square meter of roost surface. Despite this, female free-tailed bats are usually able to locate and feed their own off-spring, with mistakes in nursing only happening around 17 percent of the time. How are bat mothers able to successfully locate their own pups in the chaos?

Mother free-tailed bats don't roost with their pups, instead spending their day in a different area of the roost but visiting their pups several times a day. Even when flightless, bat pups are still mobile and move around within the creche between nursing visits. When they come to nurse, females land on or near the creche and then start crawling through the cluster of pups. As they

 A female Mexican free-tailed bat looks for her young in a crowded creche.

search for their young, females pause and produce repetitive directive calls accompanied by rapid ear flickering as they listen for a response. These directive calls vary across individuals, potentially enough for pups to recognize the calls of their mothers. Pups produce their own calls while in the creche, termed isolation calls, which are described as "peeps," with a combination of human-audible and ultrasonic frequencies that also show individual variations. Hungry pups are generally less discriminating than females, approaching and trying to nurse from any passing adult, while the females try to repel these milk thieves by beating and scratching at them with their wings and feet.

Once they are close enough to touch a potential pup, mother free-tailed bats use odor cues to discriminate their own offspring from other pups. In experiments, when presented with the odor of their own pup and the odor of a randomly selected pup, lactating females correctly chose the smell of their pup. Interestingly, pups did not show a preference for the odor of their mother, choosing the odor of a random female over their mother's about half the time.

Before accepting a pup to nurse in the creche, a female investigates the pup with her muzzle, smelling and vocalizing. However, even with the potential to recognize individual pup's calls or smells, cave roosts are a complex environment with a lot of potentially interfering acoustic and olfactory stimuli.

To narrow things down, female bats appear able to remember the rough location of where they last nursed their pup. When researchers temporarily moved pups from a specific area of the creche, females started their search where they last saw their pup, suggesting that they remembered the spatial location in the roost. There are also times when moms may not be able to locate their own young, either due to crowded conditions or fallen pups. The floor of caves like Bracken Cave can be rich ecosystems, full of invertebrates and other critters, all subsidized by bat guano and carcasses of unlucky bats—including bat pups.

Evening bat pups also produce isolation calls when females return to the roost after foraging, but females of this species seem to rely only on olfactory cues to confirm recognition of their own offspring. Olfactory cues also seem to be part of the offspring search behavior in long-eared bats and pipistrelles. Overall, very little research has directly tested the role of olfactory cues for communication in bats, including between mother and pups.

On the other hand, the use of acoustic cues between mothers and pups is better understood. Many species use contact or isolation calls, including non-echolocating species like flying foxes. These isolation calls may have both a genetic and learned component. Following birth, the pups of many species vocalize almost continuously, which might maximize the ability for females to learn the sounds of their own pups. In spear-nosed bats, as pups grow their isolation calls gradually change to more closely resemble the calls of their mother, likely also improving mother-pup recognition. When spear-nosed pups were experimentally isolated from other bats and hand-raised, only those that were exposed to playback of their mother's directive calls showed this pattern, confirming that pups are able to learn and mimic the calls of their mothers.

# Sharing the Burden

What we know from mother-pup interactions indicates that females are generally good at locating their pups, even in crowed spaces, but mistakes do still happen. From an evolutionary perspective, expending energy to nurse and care for a pup that is not your own is disadvantageous, as doing so does not ensure survival of your own genes. Despite this, communal care occurs in a variety of mammals, including humans. Driven by kin selection, this often involves the care of relatives, where the caregiver still shares some genetic material with the receiver. Given the potentially complex social lives of bats and the vulnerability of young pups, do bats ever care for offspring that aren't their own?

Communal nursing has been observed in natural roosts of evening bats, though still at relatively low frequencies (about 18 percent of observed nursing bats in one study). In this case, kin selection did not seem to be a factor, as the likelihood of a mother nursing a pup other than her own was not predicted by relatedness. Instead, such behavior was observed more often close to the end of the nursing period (2 weeks from weaning). Proposed explanations include milk-dumping by some females to reduce their weight before foraging or potential tit-for-tat mutualism like what is seen in vampire bat blood-meal donations.

Even without being related, cooperation and communal care can still be beneficial to individuals living in stable social groups. Greater spear-nosed bats in Trinidad exhibit the unusual behavior of pup guarding. In these colonies, pups falling from the cave ceiling to the floor is a major contributor to pup mortality. Adults will visit fallen pups, with some individuals retrieving or capturing the pups. In other instances, adults will instead attack or bite the fallen pups. Females in this species form small stable groups that forage together and give birth once a year around the same time; pup mortality is high, with about 25 percent of pups failing to survive to 6 weeks. Under these conditions, any behavior that increases pup survival can have a positive effect on reproductive success. Researchers simulated a scenario where a pup had fallen by placing pups low on the cave wall and then observed how adults interacted with them. Females were more likely to visit fallen pups from their own social group, and this behavior may reduce the likelihood of pup attacks from other adults.

An extreme example of nonkin care of offspring in bats was recently observed in captive common vampire bats via adoption. Wild-caught vampire bats were being kept in captivity at the Smithsonian Tropical Research Institute as part of ongoing studies on cooperation. After being brought into captivity, two female vampire bats developed a food-sharing and grooming relationship. After a little while, one of the two gave birth to a female pup, and the social partner extended her grooming and food sharing to the pup as well. Shortly after giving birth, the mother bat died unexpectedly, at which point the social partner—who had not given birth—began lactating and providing care to the 2-week-old pup. While remarkable, it is important to note that this adoption happened under captive conditions, and it is unclear how likely this behavior is to occur in the wild.

◄ A long-tongued bat of the genus *Glossophaga* leaves the roost with her pup clinging to her belly.

# On the Move

Sometimes, between a pup being born and learning to fly, the pup has to be moved. In some species, pups remain attached to mom for an extended period of time, even when she is flying and foraging. Transport of young bats by their mothers during foraging is mostly observed in fruit- and nectar-feeding bats and only rarely in insect-feeding bats. Carrying young increases the energetic demands of flight, reducing maneuverability and foraging efficiency.

Carried pups usually wrap themselves around the mother's belly, latch onto a nipple, and cling with their developing wings and feet. Bats that roost in ephemeral, temporary roosts such as leaf tents also move their pups throughout the nursing and weaning period by carrying them. Mother bats will move pups between tree roosts or areas of a given roost, depending on factors like weather and thermal environment.

In addition to the usual set of nipples for nursing, some bats have an extra nipple or two located in the hip area. Referred to as pubic or holdfast nipples, this feature is seen in a range of bat families, including female bumblebee bats (Craseonycteridae), false vampire bats (Megadermatidae), mouse-tailed bats (Rhinopomatidae), and horseshoe bats (Rhinolophidae). These nipples appear to function as an extra latch point for young bats during transport. A pup clings to its mother belly-to-belly, holding onto the pubic nipple with its mouth and wrapping its legs around mom's shoulders. Amazingly, in the greater mouse-tailed bat, diadem roundleaf bat, and little Japanese horseshoe bat, these pubic nipples contain duct systems and lacteal tissue. Thus, it appears that in these species the "false" nipples may actually function to nurse pups, though more work is needed to confirm this behavior.

# Bat Dads

For the most part, male bats are not involved in raising pups. In temperate bat species, where females form maternity colonies to give birth and raise young, males roost in separate locations. In some species, males roost together in bachelor roosts, whereas in others each male roosts alone. In tropical bat species that roost in small harems, with a single male and multiple females, males continue to share the same roost, but they don't contribute directly to pup care.

There are a few potential exceptions to this rule. Large, Neotropical carnivorous bats like the spectral bat and the woolly false vampire bat appear to live in small family groups. A roost of spectral bats in Costa Rica was composed of two reproductive adults, two late-stage juveniles, and one nursing pup. Both adults were seen bringing prey back to the roost, suggesting both mom and dad were engaging in parental care.

Male bats also contribute to the creation and defense of stable roosts. For some species that make tents out of leaves, most of the tent construction is done by the male. Male short-nosed fruit bats defend both these tents and the female occupants throughout pregnancy and pup rearing, thus providing a relatively safe place to raise young. The same is true for the termite-mound-chewing males of the round-eared bat, who devote a significant amount of time to carving out space inside termite mounds for females and their pups to roost.

The most unusual find related to paternal care was the discovery of lactating male bats. Researchers studying bats in Borneo in the early 1990s were surprised when they captured several male Dayak fruit bats who were leaking what appeared to be milk from their nipples. Later dissection of the mammary glands of some of these bats revealed lactiferous (milk-associated) ducts, like those found in nursing female bats. Does that mean these male bats were nursing young? Probably not. For one, the amount of milk produced by the captured

▼ The spectral bat is the largest bat in the western hemisphere. Carnivorous species like this one are often found living in small family groups, and both parents are thought to contribute to pup care.

males was very small, only about 2 percent of the volume of a lactating female. Second, nursing pups leave distinctive evidence of gnawing and chewing on nipples, which was not observed on these captured males.

Milk production by males is not unheard of in other mammals. In the medical sciences, milky discharge from male nipples is called galactorrhea and has been observed in domestic livestock and humans. It is usually temporary, caused by some disturbance in hormone levels due to certain diseases, diet, or drug side effects. In bats, observations of milk-producing males have been limited; none of the Dayak fruit bat males captured in the same region in a different year showed evidence of lactation. Exactly why this occurred in these bats is still a mystery. Perhaps it was a diet high in plant estrogens or some malfunction or disease of the liver that affected hormone regulation. Either way, Dayak fruit bats and another bat species, the Bismarck masked flying fox, remain the only known cases of male lactation by bats in the wild.

# Babbling Baby Bats

Learning to speak and echolocate takes time, and there are some surprising similarities between humans and bats when it comes to developing their voice. For bats that echolocate using their vocal cords (laryngeal echolocators), the ability to produce and process echolocation calls appears to be closely linked to the timing of flight.

▲ A 3-day-old big brown bat pup practices using its voice.

Although not able to echolocate, nearly all newborn bat pups are able to produce variations of isolation or directive calls. Pup isolation calls usually have a lower frequency and longer duration than adult echolocation calls. Some bat pups are unable to hear high-frequency calls, and other species are born completely deaf. For example, newborn short-tailed fruit bats are only able to hear about 68 percent of the adult hearing frequency range and are most sensitive to lower frequency sounds. Similar patterns have been recorded in the pale spearnosed bat and mustached bats.

As pups age, the variety and frequency of their vocalizations increase. Vocalizations with frequency patterns and harmonics that resemble adult echolocation calls start somewhere between the first and third week of life, depending on the species. In most species, by the time pups have reached the age of flight, their echolocation calls are nearly indistinguishable from that of adults.

During this period of learning to vocalize, baby bats produce a range of different sounds, like the way human babies babble. This has been studied most closely in big brown bats and sac-winged bats. Sac-winged pups were found to produce a range of different syllables, sometimes repeating the same syllables over and over. Importantly, these repetitive vocalizations were not associated with any specific behaviors, the way that pup isolation calls are produced to attract a mother's attention. A similar, though not as extensive, behavior was also observed in big brown bats.

The development of bat echolocation is partially explained by growth and development in the rest of the body. Like many other species, Jamaican fruit-eating bat pups produce isolation calls starting right after birth before gradually developing long, sweeping vocalizations as they age. The pattern of increase in frequency and call length appears to be linked to growth of the larynx. As pups age, the cartilage in their larynx becomes hard and bony and more like that of adults. Their cricothyroid muscle, one of the muscles responsible for helping to produce movement in the vocal cords, also grows larger and stronger as pups age. Although young bats can produce high-frequency calls without this calcified cartilage, ossification may make it easier to produce high-intensity (loud) calls in bats. The timing of this larynx development also coincides with growth in wing and flight muscles that enable bats to fly on their own.

Some of the earliest development of echolocation occurs in the click-echolocating Egyptian fruit bat. Newborn pups in this species are not only able to hear and produce high-frequency clicks, but they produce them in pairs and shape the sonar beam almost like adults do. As they age, Egyptian fruit bat pups increase how fast they produce the click pairs, reaching calls identical to adults by the time they begin to fly.

# Flight School

Learning to be a bat means more than just learning to echolocate—it means learning to fly! All bats follow the same general sequence for learning to fly, the four Fs. The first phase occurs immediately after birth, as baby bats *flop* when gently dropped from a short height. They make no attempt to open or flap their wings and simply fall straight down. This phase tends to be short, lasting between 5 and 10 days in most species. As the pups grow, they enter their *flutter* phase, when they move their wings as they fall but still fall straight down with no movement forward. Third is the *flap* phase, when juvenile bats become capable of horizontal movement while flapping, though they can only move a few meters before hitting the ground. Finally comes *flight*, when young bats can stay airborne in a sustained and controlled manner.

Some of the most detailed studies on the development of flight in bats have focused on temperate, insect-eating species like the big brown bat and the little brown bat. Juvenile bats in these species start flying as early as 4 weeks after birth and begin hunting on their own at 6 to 8 weeks of age. Larger, fruit-eating species, like short-tailed fruit bats and Jamaican fruit-eating bats, take a little longer to reach full flight status: around 30 and 35 days of age, respectively. Although juveniles fruit bats are able to maintain sustained flight at these ages, they do not reach adult maneuverability in tight spaces until closer to 2 months of age.

# Bite School

You're a juvenile little brown bat, having just hit 6 weeks of age. You've got those flight muscles working hard and the echolocation calls dialed in just right. It's time to explore the wide-open skies! And maybe give your hard-working mom a break after many intense weeks of carrying and nursing you, right?

 Yellow-winged bats learn to hunt by following their parents to feeding territories.

While bat moms might start to get a little bit of a break, maternal care doesn't stop after pups become airborne. Weaning bats off mother's milk also occurs around the same time as flight, ending as juveniles become proficient at capturing their own food. Predictably, juvenile body mass decreases slightly as bats learn to fly as they are using more energy each day. For some species, young bats may learn to hunt from their mothers, at the very least following them on foraging flights. Juvenile yellow-winged bats, which glean insects and other prey off the ground by listening from perches, accompany their parents to feeding territories and even share their parents' foraging perches. Mothers of another gleaning bat, the common big-eared bat, bring insects to their pups even after juveniles start flying and hunting on their own. This maternal provisioning is thought to help these young bats learn to handle large and potentially unwieldy prey items.

Following mom doesn't seem to occur in all species, with some juvenile bats showing evidence of learning to hunt independently. Using artificial flowers equipped with radio frequency identification tags, researchers in Costa Rica were able to track the feeding patterns of individual long-tongued bat mothers and pups. The mothers and pups rarely visited the same flowers, with pups visiting flowers closer to the roost than their mothers. These findings suggest that juvenile long-tongued bats don't follow mom to flower patches and instead explore their environment on their own. While mothers might not be helping away from the roost, unlucky or unsuccessful pups can still count on mom. Long-tongued bat mothers continue to nurse their young even after they can fly and will provision subadult bats with regurgitated nectar.

In Egyptian fruit-eating bats, teaching young bats good hunting areas and how to get around starts before they learn to fly. Pups stay attached to mom for the first 3 weeks after birth, even as she flies around and feeds at fruiting trees. As pups get older, mom will start leaving pups at trees away from the roost. Even as pups gain independence and the ability to fly, the mother bat continues to bring them to these trees. From these drop-off trees, juvenile bats begin making small exploratory flights and finding their own fruiting trees. By bringing pups with them and then leaving them at these temporary stops, the mothers help their offspring learn the area around the roost.

Maternal guidance may also be important to help young bats find roosts, especially in species that switch roosts throughout the year. During the summer, noctule bats form maternity colonies where they give birth and rear their young. These colonies are fluid, with individuals moving between several roosts, and noctule bat pups continue to nurse from their mothers even after fledging. Using small proximity sensors that recorded when two bats were within a short distance of each other, researchers in Germany tracked the movements of mother-pup pairs during foraging and roost movements. Although mothers and pups did not tend to be near each other during foraging bouts, juvenile bats were often near mom during roost-switching events. In some cases, mothers would pause at stopover sites, waiting for their young to catch up before herding them toward the new roost.

Parental investment continues beyond weaning and fledging in many bats, but sometimes maternal instinct results in a good strong shove out of the roost. While using infrared video to observe juvenile growth in Peter's tent-making bat, researchers noticed some strange and repetitive behavior. Beginning when pups were about 25 days old, mothers would start tapping and nudging their attached babies with their forearms shortly before heading out to hunt. During this prodding, pups would detach from the mother and then crawl back over and try to reattach. As pups grew older, mothers didn't need to spend as much time elbowing their young to get them to move. By about 40 days of age, young pups were ready to spread their wings on their own and take flight.

# SPILLOVER

## Chapter II

Three hundred eighty trillion.

That's how many virus particles are estimated to be living on and inside your body right now, as you are reading these words. The virome is the scientific term for this collection of viruses. Of course, not all of those viruses make you sick. Some might even be helpful—infecting and killing bad bacteria or viruses in your body. Only about 270 viruses are known to infect humans, representing at least 23 virus families. Of these disease-causing viruses, about one-third seem to infect only humans.

As you can see from those numbers, we humans are a source of more than a few disease-causing viruses. Chances are, however, that you've heard a newscast or read a headline exclaiming that bats carry many diseases (mostly viruses) and that they are unique among disease-carrying animals. These often-sensationalized headlines have become more common in the past few years as we've collectively grappled with the emergence of and ongoing consequences of the COVID-19 pandemic. But are bats really that special when it comes to diseases? And what kind of role did they play in the emergence of SARS-CoV-2, the virus responsible for COVID-19?

▶ Flying foxes glide across the night sky in Sri Lanka.

# Viruses and Reservoirs

Viruses consist of a packet of genetic material (either DNA or RNA) encased in a protein coat. Viruses are not able to replicate on their own. Instead, they infect cells and hijack the machinery of the host cell to make copies of themselves, destroying the host cell in the process. Viruses infect cells from across the tree of life, from single-celled bacteria to plants and animals. While they might sound simple, viruses show an incredible diversity of shapes, types of protein coats, sizes, and complex structures on their coat that help them to invade cells while evading the host immune system.

Infectious microbes, including viruses, bacteria, parasites, and pathogenic fungi, that move from a nonhuman animal to a human are referred to as zoonotic diseases (or zoonoses). Of the 200+ disease-causing viruses found in humans, about two-thirds are zoonotic, meaning that at some point in the viruses' evolutionary history they were transmitted from a nonhuman animal to a human. How a disease jumps from one species to another varies, with transmission happening via direct contact (such as a bite or from petting or handling sick animals), consumption of infected and untreated meat or other animal by-products (like milk), or contamination from animal excretions. The host in which a given pathogen (disease-causing agent) naturally lives and reproduces is referred to as the animal reservoir. Reservoirs can be humans, other animals, or even just the surrounding environment. Animal reservoirs of a virus or bacteria can—but often do not—experience symptoms of that disease.

For many zoonotic viruses, spillover doesn't happen directly between the reservoir species and humans. Instead, the virus is first transmitted to an intermediate host, where it may multiply or acquire new adaptations before infecting humans. Common intermediate hosts include pigs, horses, camels, and other domestic animals. So-called spillover events between animals may happen more often than we realize, but most end up reaching a dead end, with no subsequent transmission. Spillover events are recognized only when a virus can continue transmission among individuals of the new host, usually through some kind of mutation and natural selection. This sustained transmission is what results in disease outbreaks.

Where and how a given virus moves between species depends on a variety of factors, including how specialized the virus is, the exact mechanism by which it gains access to a cell and evades the immune system, how quickly the virus mutates as it replicates, modes of transmission (for example, whether contact is required or if the virus can spread via air particles), the biology and immune system of the host, and how that host interacts with other organisms and its environment. The evolutionary trajectory for zoonotic diseases is often convoluted and confusing. For example, the virus that causes measles in humans today is species specific, meaning that it only infects humans. But in the past some version of the measles virus infected and circulated among cattle before spilling into human populations, where it mutated and adapted to become the human-specific virus we see today.

# Finding Viruses in the Wild

As emerging zoonotic diseases become a greater concern worldwide, active surveillance of known and potentially new viruses in wildlife has increased. The goal of these surveillance programs is to randomly sample from wild and free-living animals, in an effort to minimize the potential bias associated with only testing animals that encounter humans or show signs of disease.

The most widely used method for sampling viruses from animals, including bats, entails examining blood samples for the presence of antibodies to certain viruses or bacteria. Antibodies are special proteins used by the immune system to identify and attack foreign objects in the body, like viruses. Antibodies work by identifying a part of the pathogen called an antigen, and each type of antibody binds to a specific antigen like a lock and key. The presence of an antibody in a sample can indicate that an individual was exposed to enough of that virus that the immune system mounted a response. However, antibody tests are usually not able to discriminate if the presence of antibodies is due to an active or prior infection.

Blood or tissues samples can also be tested for the presence of a specific virus using techniques like polymerase chain reaction (PCR), which allows researchers to amplify the number of copies of a virus's DNA or RNA to allow for analysis of the genetic sequences. Antigen testing, the method by which we can now test ourselves for viruses like COVID-19 at home, works in a similar way by specifically looking for the presence of a known viral antigen. The downside of many

▲ Active sampling from wild bats requires capturing random individuals using mist nets (pictured here) or harp traps. Passive surveillance occurs when sick or dead bats are collected and tested for disease.

of these molecular techniques is that they require previous knowledge of the genetic sequences of a pathogen and are often species specific.

Newer technology such as next-generation sequencing has made it possible to screen for unknown viruses in wildlife. In a multistep process, next-generation sequencing generates DNA or RNA sequences that can be compared to the sequences of known genes or viruses. Sequence similarity is reported in percentages, but it's important to note that similar sequences do not necessarily mean similar biology. For example, humans and chimpanzees are estimated to share about 98 percent of their DNA sequence but are clearly very different physically, behaviorally, and ecologically. Viruses also mutate and evolve quickly, so just because a virus originated in a bat doesn't mean that bats carry the current version of the virus.

# A Brief Summary of Bat-Associated Diseases

Much of the recent discussion and coverage of bats and diseases focuses on viruses, though there are a few other bat-associated pathogens that have the potential to impact human health. One of the most well-known nonviral diseases associated with bats is histoplasmosis, an infection of the lungs caused by breathing in spores of the microscopic fungus *Histoplasma capsulatum*. The fungus lives in soil, with human exposure usually happening after activities like construction that disrupts soil or spending time in bat caves or bird roosts without protective respiratory equipment like masks. Despite living in conditions with constant exposure to the fungus, there is no evidence that bats experience negative effects from *H. capsulatum*.

Like all living things, bats can also act as viral reservoirs. About 12,000 bat-associated viral sequences from 30 viral families have been identified in bats, though not all these viruses are necessarily zoonotic or unique. Both DNA and RNA viruses have been identified in bats, with RNA viruses making up the largest proportion. Families of RNA viruses most commonly detected in bats include coronaviruses, rhabdoviruses (rabies), paramyxoviruses (measles and mumps), and astroviruses. Currently, none of the astroviruses found in bats are known to be associated with disease in humans. Other notable viruses with known impacts on human health that have been associated with bats include those in the family *Filoviridae* (specifically *Ebolavirus* and *Marburgvirus*).

## Bats and rabies

Teeth bared, mouth foaming with saliva, loud ferocious growling. That's the picture that's frequently associated with rabies, mostly thanks to pop culture depictions and horror movies. Despite being a vaccine-preventable disease, rabies causes an estimated 60,000 deaths across the world every year and is considered a neglected tropical disease by the World Health Organization. Although 99 percent of global rabies transmission to humans are from domestic dogs, bats are still the first animals implicated in pop culture discussions of rabies, which has led to a lot of confusion and misinformation about bats and the disease.

*Lyssavirus rabies* (or RABV) is an RNA-based virus in the family *Rhabdoviridae* that affects the nervous system. It is transmitted when saliva or nervous system tissues (such as brain) of an infected animal contacts the broken skin or mucous membranes (such as in the eyes or mouth) of another animal. The most common mode of transmission is a bite from an infected animal, although RABV can sometimes be transmitted via scratches or abrasions and even more rarely via aerosols. The incubation period, or how long between exposure and the start of symptoms, varies widely from just a few weeks to months. Late-stage clinical symptoms come in two forms. The most recognizable is furious rabies, which is characterized by hallucinations, delirium, fear of water (hydrophobia), and insomnia. The second is paralytic rabies (also known as dumb rabies), which is characterized by weakness, paralysis, and even coma. Once the disease takes hold, there is no cure for rabies. However, rabies is highly preventable with access to vaccines and can be treated if the victim quickly receives a series of shots that prevent the virus from infecting the body.

Can bats carry rabies? Yes. Almost every mammal is susceptible to the rabies virus, and bats are no exception. There are several variants of RABV, some of which circulate in bats and others that are found in mammals like skunks, raccoons, and canids (dogs). Bats become infected with rabies the same way as other mammals, most likely from bites by other bats. There are occasional reports of bats attacking and even trying to consume each other. One report published in 1947 described an aggressive attack between a hoary bat and a tricolored bat encountered by a couple on their nightly walk. I witnessed an aggressive incident between two bats while doing mist-net surveys in northern California. After hearing bat distress calls, it appeared that a hoary bat had attacked a silver-haired bat, forcing them both to the ground, with both bats sustaining minor bite injuries. We didn't test either bat for rabies, so we don't know what caused this fight, but similar (though rare) observations have been associated with rabies-positive bats.

Bat variants of RABV are less common outside of the western hemisphere. Of the seven genotypes of RABV, three are found in Europe and Asia. Genotype 1 (the one we usually think about when we think of rabies) is thought to have evolved and spread in canids within Europe before spreading to Asia, Africa, and the Americas. Most European countries are considered rabies-free after having eradicated this variant, though several closely related rabieslike viruses have been isolated in bats (European lyssavirus type 1 and type 2). In addition to RABV, there are at least sixteen other *Lyssavirus* species globally, most of which are associated with bats as reservoirs. For example, Australian bat lyssavirus was first isolated in the black flying fox but has also been detected in a few species of fruit- and insect-eating bats. There are almost no reports of non-RABV bat lyssaviruses infecting humans.

Do *all* bats carry rabies? Not even close. In studies where North American insect-eating bats were randomly surveyed for rabies infection, prevalence averaged less than 1 percent. The prevalence of rabies-virus variants in Europe is similar, estimated at about 2 percent.

A common myth about bats and rabies is that they can carry and spread the disease without experiencing symptoms, but this is untrue. Some bats may start to shed the virus up to 2 weeks before showing clinical symptoms, at which point the disease progresses like normal until the animal dies. What is notable, however, is that bats are very good at fighting off rabies. Mammals like dogs and humans need to be exposed to an inactivated version of the virus through vaccines to gain immunity, whereas bats are able to produce their own anti-RABV antibodies after natural infection, stopping the disease from progressing further and protecting them against future infection. In 2021, the most recent year for which data is available, the primary species with confirmed rabies infections were bats, raccoons, skunks, and foxes (in that order). Of the roughly 22,000 bats submitted for testing, only about 5 percent tested positive. In contrast, about 20 percent of the skunks and foxes submitted were rabid. This type of passive surveillance can result in misleading statistics, however, as sample submission is usually triggered by some unusual event, such as strange behavior or human exposure via a bite.

Do bats cause rabies in humans? Rarely. The incidence of human cases of rabies in the United States and Canada is extremely low, with 127 human rabies cases reported in the United States between 1960 and 2018, about 65 of which were attributed to bats. In the United States, death from rabies in humans is almost always due to a lack of postexposure treatment, which is extremely effective. In Central and South America, rabies from bats accounts for about 40 percent of human cases—although total number of reported cases is still low, with 192 cases reported across 7 countries over 10 years.

Given that other mammals that also frequent areas of human activity, like skunks and raccoons, are also reservoirs of rabies, why are bats most often associated with fatal cases of rabies in humans? It's because relatively few rabid bats show the easily recognizable signs of furious rabies, with most showing symptoms like paralysis and loss of coordination, which can make them seem tame, calm, and less of a threat. Bat bites can also be very small and easy to overlook as trivial injuries (or not noticed at all). Unrecognized or disregarded bites can mean an exposed person is less likely to seek medical treatment and receive postexposure vaccination. Current guidelines in the United States also recommend postexposure vaccination in cases where it is unknown if a bite might have occurred, such as finding a bat in a room with unattended children or a sleeping person. This is not because being in the same space as a bat is inherently dangerous, but rather to account for potentially unknown contact or exposure.

▲ Bats that roost in human structures (like these Yuma myotis roosting in an attic) are more likely to be encountered by humans, even when the bats are healthy. Never handle a bat with your bare hands, and consult your local wildlife rehabilitator for what to do if you find a bat in your house.

# Disease Outbreaks and Bats

While RABV is confirmed to transmit directly between bats and humans, humans are a dead-end host for the rabies virus. In the last 100 years or so, only three virus families have successfully crossed species barriers and resulted in human outbreaks: filoviruses, henipaviruses, and coronaviruses. The strength of the association between these groups of viruses and bats is variable and, despite decades of research in some cases, still somewhat speculative.

# Filoviruses: Marburg and Ebola

Although rarer than many other types of viruses, filoviruses are of particular concern to human health due to their infectiousness and high mortality rate in those infected. The most well-known examples are Ebola virus (EBOV) and the virus causing Marburg disease (MARV). Both filoviruses cause severe, often fatal illnesses that have resulted in outbreaks in various parts of the world, mostly in Africa.

In the decades after the first known outbreak of Ebola in 1976, EBOV had only been identified in people and nonhuman primates. EBOV-specific antibodies were not isolated from bats until 2003 and have since been detected in some populations of African fruit bats, including straw-colored fruit bats, hammer-headed bats, and Egyptian fruit bats. EBOV antibodies indicate that bats have been exposed to EBOV and able to mount a sufficient immune response, characteristics that indicate a reservoir host. Although genetic sequencing has found viruses similar to Ebola in some bats, live EBOV has never been isolated from a bat, despite extensive sampling efforts. There is also limited evidence of Ebola emerging following exposure to bats, with only two instances of patient zero in an outbreak having some form of known bat exposure (indirect exposure to bat meat in one case and playing near a known bat roost in the other). In one of these cases, later analysis of the suspected patient zero's activities suggested that exposure may have come from another person who was experiencing symptoms from what is now recognized as post-Ebola syndrome. A flare-up of viral activity and symptoms in survivors of Ebola, post-Ebola syndrome could explain more recent outbreaks without there needing to be a specific spillover event from wildlife.

Both MARV and MARV antibody isolates have been identified in at least one species of bat, the Egyptian fruit bat. MARV antibodies have also been detected in other fruit bats and one insectivorous species. In surveillance following an outbreak in Ugandan miners in 2007, the gene sequences of MARV isolated from bats and from infected workers were closely matched, suggesting either bats were the source of the infection or both the workers and bats were exposed to the same source within the cave system.

Despite decades of research looking for links between bats and filoviruses, there is still a lot of uncertainty about the role of bats in human cases. The prevalence of filovirus-associated antibodies or disease cases in wild bat populations

appears to be low (2–12 percent). Potential filovirus spillover events are still rare, although the risk may rise with an increase in activities like entering bat roosts and caves. Definitively stating that bats are the natural reservoirs of viruses similar to Ebola can be problematic, as it may encourage fear and persecution of these ecologically important animals.

▲ Straw-colored fruit bats are a widespread fruit bat in Africa that is commonly found near human activity. There is still no definitive proof that fruit bats are the main drivers of Ebola in humans.

## Henipaviruses: Hendra and Nipah

The family *Paramyxoviridae* is a large and diverse family of RNA viruses that causes a range of human and livestock diseases, including measles, mumps, and distemper. While bats as a clade appear to host a range of viruses in this family, the most well studied are in the genus *Henipavirus*: Hendra virus and Nipah virus. In Australia, flying foxes in the genus *Pteropus* are thought to be the primary natural reservoirs for Hendra virus. There have been very few cases of Hendra virus in humans and no evidence of direct transmission from bats. The intermediate hosts are domestic horses, which can be infected after exposure to the urine of infected flying foxes.

Nipah virus was first recognized during an outbreak in Malaysia in 1998, and others followed in Bangladesh and India. Like Hendra virus, the primary natural reservoir of the virus appears to be *Pteropus* bats. Pigs are the proposed intermediate host between bats and humans, although some transmission has occurred between bats and humans via palm sap contaminated with infected bat saliva. Antibodies to Nipah virus have been detected in a wider geographic range than those of Hendra virus, including in bats in Cambodia, Thailand, China, Madagascar, and Ghana, though no human cases have been reported in these locations.

## Coronaviruses

Since the spread of COVID-19 on a global scale in early 2020, coronaviruses have taken center stage when it comes to viral zoonotic transmission. Coronaviruses have been known in humans and other animals for a long time. A family of RNA viruses, members of *Coronaviridae* have large genomes and are known for their rapid mutation rates that let them quickly adapt to new hosts and environments. Coronaviruses are divided into four genera, with *Alphacoronavirus* (alpha-CoV) and *Betacoronavirus* (beta-CoV) mainly infecting mammals. Both can be found in bats, but they also infect carnivores, lagomorphs (rabbits and hares), primates and humans, hooved animals, and rodents.

In 2002, severe acute respiratory syndrome, caused by a new beta-CoV (SARS-CoV), emerged as the first global epidemic caused by a coronavirus. Prior to its discovery, a handful of coronaviruses were mostly known to cause disease in domestic animals and humans, including several varieties of the common cold. It was the search for the origin of SARS-CoV that first linked coronaviruses and bats. Early evidence based on antibodies and PCR testing suggested that a few species of horseshoe bats (genus *Rhinolophus*) may be natural reservoirs for the virus. It wasn't until 10 years later that researchers were able to isolate a bat-associated coronavirus with a high genetic similarity (about 95 percent) to the SARS-CoV that infected humans, suggesting an ancestral origin in bats.

◄ Horseshoe bats (genus *Rhinolophus*) are a diverse group found throughout Asia, Africa, and Europe that may serve as a natural reservoir for SARS-CoV.

Since then, viral surveillance efforts have identified more than 4000 coronavirus sequences from bats, though these are not necessarily all separate virus species or capable of infecting other animals. This high diversity has led scientists to hypothesize that ancient strains of coronaviruses may have originated in bats sometime in the 50+ million years that bats have existed, eventually diversifying into other mammals.

In 2012, Middle East respiratory syndrome (caused by MERS-CoV) emerged as another severe coronavirus disease affecting humans. Human cases of MERS-CoV were transmitted from dromedary camels, and humans appear to be a dead-end host for the virus, with very little human-to-human transition. A handful of genetically similar MERS-like coronaviruses have been isolated in bats, with the closest virus in a bat having about an 85 percent sequence similarity to MERS-CoV. Although MERS-CoV might have had its ancestral origins in a bat, in terms of risk to humans it is considered a camel virus, not a bat one.

Then came 2020 and the start of the COVID-19 pandemic. After early clinical studies indicated this new respiratory disease was caused by a SARS-like virus, one of the big questions was where it came from. Thanks to the intensive viral surveillance of bats and other wild animals in the years following the discovery of SARS-CoV, many SARS-like coronaviruses had already been sequenced and could be used for comparison to the new human SARS-CoV-2. From these samples, the closest match was RaTG13, a coronavirus sequenced from bat samples collected in Yunnan Province, China, sharing 96.2 percent of its gene sequence with SARS-CoV-2. By looking at possible changes in gene sequences and known mutation rates between SARS-CoV-2, RaTG13, and other similar coronaviruses, scientists estimate that known bat viruses and the coronavirus causing COVID-19 actually diverged between 40 and 70 years ago. While not definitive, it suggests that the current SARS-CoV-2 outbreak may not be the result of recent bat-human exposure.

So where does that leave the origins of SARS-CoV-2? Still uncertain. As we saw with the emergence of new variants of SARS-CoV-2 in humans after the first wave of infections, this virus is quick to adapt and change. While the prevailing scientific hypothesis for how this virus moved into humans is still centered around zoonotic transmission, we may never find the original source. What is important is that the spread and continued persistence of the virus is all driven by human activity, and bats are not to blame.

## Are Bats Unique When It Comes to Diseases?

Over the years, there has been much speculation that bats may be unique in their potential to act as reservoirs of zoonotic diseases, fueled by already described outbreaks and increased surveillance for wildlife diseases. Unfortunately, the answer to this question is not a simple yes or no, but rather, it all depends on your definition of *unique*.

Earth is home to a lot of bat species: 1474 and counting. With so many species, we might expect there to be a lot of bat viruses. After all, the single species *Homo sapiens sapiens* (that's us humans) can have over 200 disease-causing viruses—and that's not even counting viruses that might be circulating in our bodies without making us sick. So, how do bats as a group compare to other mammals when it comes to viruses?

It depends on how you compare viruses. In one frequently referenced study from 2013 that compiled all the known viruses in bats and rodents, bats were found to contain slightly more viruses on average than rodents (2.71 viruses

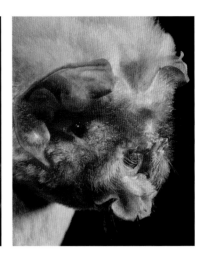

per bat species compared to 2.48 viruses per rodent species). A 2017 study also reported that bats host a higher proportion of zoonoses when compared to other mammals. However, a more recent analysis published in 2020 suggests that the number of human-infecting viruses found in bats is all relative. When accounting for the diversity of bats as an order, bat viruses are not any more likely to be zoonotic than viruses found in other mammalian reservoirs. The number of viruses detected in bats is proportional to the number of species.

The types of viruses may also matter. Bats tend to have more RNA-based viruses, which coincidentally make up the majority of known zoonotic viruses. Factors like how closely related an animal reservoir is to humans and exactly how and when humans encounter that reservoir also matters. A study published in early 2022, led by Sara Guth at the University of California, Berkeley, examined differences in disease virulence (how severe a virus infection is) and a virus's capacity to spread within the human population (transmissibility) in a set of mammalian and avian zoonotic viruses. Viruses from reservoir hosts that are closely related to humans, such as nonhuman primates, tended to be more transmissible but less virulent. In this study, bats were more likely to be reservoirs of virulent viruses, but the authors also found that the death burden (or estimate of human death from a given zoonotic virus) was not associated with any given animal reservoir, including bats. Put another way, this means that where a zoonotic virus comes from is not predictive of how dangerous it might be, and every potential viral outbreak is unique.

▲ Although it's easy to lump all bats together, the southeastern myotis (left), wrinkle-faced bat (middle), and ghost-faced bat (right) highlight the order's rich diversity of form and function.

Scientists may know about more bat viruses because we spend a lot of time looking for viruses in bats. Bat-virus studies have grown dramatically over the past 15 years, with many more studies focused specifically on bats than other mammalian groups. Reporting on bats and viruses often trend toward sensationalism. In the case of COVID-19, one meta-analysis found that research studies focused on bats and camels were referenced more often than studies of other animals, despite the presence of coronaviruses in many other animal groups. While many of the studies (including those referenced here) do account for sampling bias, the nuances of these methods are not always well reported in mainstream media.

To better mitigate future zoonotic outbreaks, disease biologists have also started asking if species traits predict their potential spillover risk. Traits that have been examined include geographic range and habitat preference, as well as life history traits such as life span, body size, and number of offspring produced per year. In bats, the number of viruses was positively correlated with life span, geographic range, and social group size. Living longer and being surrounded by more individuals may increase potential exposure to viruses across a bat's life span. Bats are also highly mobile, which can be a factor in how viruses can move across the landscape.

Bats also have exceptional immune systems. Although it has been argued that this makes bats special as viral reservoirs because they can carry and move these viruses without showing symptoms themselves, proving this to be true or false is very difficult. For one, it is unlikely that scientists can know all possible causes of death in a bat population. Second, it is nearly impossible to determine the health history of a wild-caught bat. The presence of antibodies and improved immune response can tell us that a bat survived an exposure or infection, but not how severe the infection was or if the bat exhibited symptoms. The only way to test the hypothesis that bats don't get sick from carrying these viruses is via experimental infection of bats in the laboratory. Results from these types of experiments are mixed, with some bats showing more symptoms than others depending on the specifics of the experiment. Understanding the intricacies of how the bat immune system can respond to these different types of pathogens and why certain pathogens might trigger a response over others also has potential to inform a broader understanding of immunity in all mammals, including humans.

◄ An eastern red bat swoops down to get a drink.

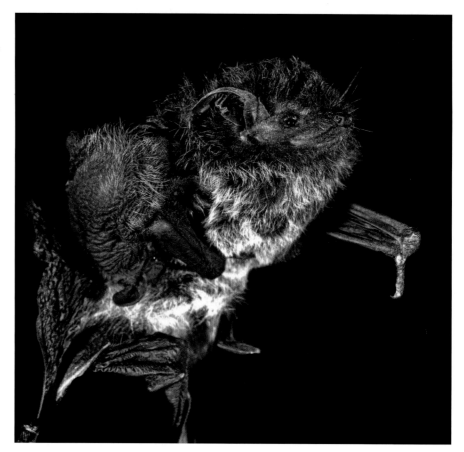

► A mother silver-tipped myotis and her newborn pup. Bats are long-lived and reproduce slowly, life history traits that may also be linked to well-developed immune systems.

# Bats and Diseases: The Take-Home Message

Disease ecology is a complex and nuanced branch of science that tries to untangle relationships between the environment, species, individuals, and, in the case of zoonotic diseases, people. Unfortunately, this nuance is difficult to capture in those flashy headlines exclaiming bats' hypothesized roles in a "global disease outbreak" or the presence of "deadly viruses" in bats. It might seem trivial, but words matter. A survey of more than 13,000 people in China found that 84 percent of them misunderstood the relationship between bats and COVID-19 and that these misunderstandings led to more negative feelings about bats overall.

Many of these discussions start with broad generalizations and assumptions about bats carrying numerous diseases or their role as reservoirs because they don't die from or experience symptoms of the disease. While bats are undoubtedly special in their amazing diversity, evidence of whether bats are unique as disease reservoirs remains ambiguous. Their seeming ability to mount immune defenses against a diversity of viruses while living long lives should not be seen as a negative trait. Understanding how bats can do this may hold the key to future treatments against such diseases. And even if that isn't the case, bats are critical parts of a global ecosystem that includes humans.

That human health is closely connected to the health of animals and the environment is the principle behind the One Health approach for addressing health issues worldwide. This approach emphasizes that caring for our environment and learning to coexist with wild spaces is inextricably linked to human health and well-being. A 2012 paper found strong evidence that modern farming and intensified agricultural practices could be linked to disease emergence and amplification, possibly due to factors such as land conversion and habitat fragmentation that bring wildlife into closer contact with humans and their domestic animals. Along with expanding populations, increased urbanization, and wildlife trade for pets or meat, human activities are what drive zoonotic spillover events—not the wildlife themselves.

# LEARN

## Chapter 12

Humans have long been infatuated with the idea of living forever, reflected in our history, myths, stories, and art: tales of magic fountains that grant the drinker restoration of youth, immortal faeries and elves, even Count Dracula, the vampiric prince of darkness. There is an entire industry of cosmetics and skincare focused on anti-aging, with the global anti-aging market valued at close to $40 billion. From a health and medical standpoint, we continue to search for cures for age-related diseases, including cancer, Parkinson's disease, and Alzheimer's disease. What if I told you that we might find clues—or even answers—to these questions in the humble bat?

▶ Members of the genus *Myotis* are particularly long-lived for their size. These little brown bats might live up to 30 years, and other species of *Myotis* have been observed to live even longer.

From fighting off and surviving potentially deadly diseases to living longer than expected to having sophisticated ways to use sound to get around, bats have evolved solutions to some problems we humans are still figuring out. And while living longer is never going to be as simple as science fiction and comic books make it out to be (unfortunately, getting bitten by a bat does not magically give you super hearing or the ability to fly), we continue to step closer and closer to a science fiction–like future.

One of the ways scientists are tackling this daunting task is through the Bat1K Project. Its name references the 1000+ species of bats that live across the world, and its goal is to sequence the genome of all living bat species. By taking a closer look at the genes and genetic mechanisms behind some of bats' amazing adaptions, like disease resistance and long life spans, researchers hope to gain a better understanding of how bats have adapted to their environments and potentially uncover insights that can be applied to human health and medicine.

## Super Immunity

The science is still murky on whether bats truly carry more diseases than other animals and if those pathogen-borne diseases are larger threats to humans than pathogens transferred from other animals. But maybe instead of thinking of bats as dangerous reservoirs who might be responsible for the next pandemic—which isn't true anyway—we should be figuring out why bats are able to tolerate or even resist the viruses they do carry. By comparing genomes among various bat species and to those of other mammals, scientists are starting to gain a deeper understanding of the mammalian immune system.

In a 2013 study, researchers sequenced and compared the genomes of the distantly related black flying fox and David's myotis. They were looking for areas of the genomes that showed signs of selection. Some genes associated with DNA damage checkpoints and cell repair showed evidence of positive selection, meaning they were being maintained or increased in frequency in the population. The researchers proposed that, while these genes may have arisen as an adaptation to the additional metabolic demands and cellular wear-and-tear associated with flight, they also function to help bats fight off infection.

Flight is the one thing that truly sets bats apart from other mammals. Flying takes a lot of energy, and bats in flight can increase their metabolic rates as much as sixteen-fold over resting. In comparison, similar-sized rodents running to exhaustion only experience a sevenfold increase in metabolic rate. High metabolic rates and accompanying increases in body temperature are also a feature of a common immune function: fever. Often seen as a symptom of infection, mild fevers help boost the performance of immune cells partly by speeding up immune signaling that encourages inflammation to fight off invading pathogens. Fevers can also stress invading viruses or bacteria, weakening them and making them easier for the rest of the immune system to fend off. Based on the physiological similarities between flight and fevers and genomic evidence of a link between immune function and the evolution of flight, some researchers have proposed a flight-as-fever hypothesis to explain bat disease tolerance and resistance.

▲ The evolution of flight—along with the associated high metabolic rate and body temperature—is proposed as one reason that bats have excellent immune systems.

Related to their high body temperatures, bats also express high levels of heat-shock proteins, which are important for helping body cells survive at high temperatures and stimulating an immune response to pathogens. Another major strategy revolves around a group of special proteins called interferons. When a virus is detected in humans, interferons are released by immune cells to help stop the virus from replicating. In some bats, however, interferons are always circulating in the bloodstream, helping fight off viral infections before they can start.

In addition to a hearty antiviral response, another strategy of the bat immune system is related to managing inflammation. Although it can be inconvenient at times, mild inflammation works to promote healing by recruiting immune cells to an injury site or even targeting certain pathogens. But too much inflammation can make things worse, leading to host cell damage and immune overreactions like those seen in autoimmune disorders. One of the ways that interferons help fight off infections is by contributing to inflammation. Despite having more interferon activity, bats don't experience increased inflammation thanks to mutations in the genes that control interferon inflammation pathways. Bats have also lost several genes related to inflammation, particularly in the PYHIN gene family. One example of these lost genes is *IL36G*, a gene known to contribute to excessive inflammation in humans with autoimmune disorders. Bats also have lost genes related to so-called natural killer receptors, which also decrease inflammation by suppressing the activity of certain immune cells.

Some of the next steps will be figuring out how to integrate these genomic tricks into treatments for humans. Mice that were genetically modified to express the bat version of ASC2, a protein involved in inflammation regulation, had reduced inflammatory responses to several viruses, including SARS-CoV-2. Since inflammation is responsible for some of the symptoms and cell damage seen from viruses like SARS-CoV-2, these discoveries may help researchers develop drugs and therapies to reduce disease-related inflammation in humans.

# Bats Gonna Live Forever?

Humans have impressively long life spans. In fact, the only mammal with a maximum life span greater than that of humans is the bowhead whale, with one individual estimated to have lived just over 200 years. On the other hand, the oldest recorded bat was a male Brandt's myotis who had lived at least 41 years at the time of his last capture in 2005. While respectable, 41 years doesn't seem that impressive compared to the centuries-old whale or even to humans. But age is more than just a number. When taking body size into account, that little (7 g) Brandt's myotis lived almost ten times longer than would be expected for its body size.

The longevity quotient (LQ) is how scientists standardize life spans across different mammals. A positive relationship exists between body size and life span: the larger the mammal, the longer it is predicted to live. LQ is the ratio between the observed life span of a species and its predicted life span based on this linear relationship. Mammals with an LQ above 1.0 live longer than expected based on size, whereas those with an LQ below 1.0 have shorter-than-expected life spans. Humans have an LQ of about 4.5. Of the nineteen species of mammals with LQs higher than that of humans, eighteen of them are bats (the other is the naked mole-rat). When averaged across species, bats live about 3.5 times longer than expected based on body size, with some species having longer life spans than others. Extreme longevity (classified as an LQ greater than 4.2) has evolved at least four times across the bat phylogeny. Some of the relatively longest lived bats are horseshoe bats (genus *Rhinolophus*), long-eared bats (*Plecotus*), and the common vampire bat and mouse-eared bats (*Myotis*). Thirteen species of wild *Myotis* have been recorded living for at least 20 years, with four species having individuals who lived for more than 30 years (including Brandt's myotis).

What might be more fascinating than these unusually long life spans is that bats don't show many signs associated with aging. As organisms age, cell and system processes begin to slow down and acquire damage, leading to conditions like wrinkled skin, arthritis, and age-associated diseases like cancer. While bats are not completely immune to diseases like cancer, they don't suffer the effects as much as would be predicted based on the situation in other mammals.

▲ Brown long-eared bats are among the more long-lived species of bats relative to body size.

# Bat's guide to life

What are the batty secrets to living a longer life? One set of clues comes from looking at patterns of DNA methylation across many different species of bats. DNA methylation is a natural process in which small molecules called methyl groups bond with a molecule of DNA. Scientists can use observed rates of DNA methylation to estimate the biological age of tissues and cells, which is referred to as the epigenetic clock. Since methylation changes gene expression by turning genes on or off, it can impact all sorts of processes, including cell growth and division, immune function, aging, and even cancer development. In 2021, researchers searched for differences in the location and rate of DNA methylation across the genome of twenty-six bat species with various maximum life spans. They found that longer lived species had less change in methylation rates, meaning that the DNA was experiencing less age-related changes. They also found that shorter lived species had higher rates of DNA methylation near genes associated with cancer suppression and immunity.

Cancer is the result of uncontrolled cell division in the body, and it can be caused by genetic changes induced by aging. It turns out that some of the same genes and pathways that help bats manage inflammation in response to pathogens also help maintain cell growth and reduce tumors. One long-lived species, the mouse-eared bat, shows increased activity in tumor-suppressing microRNAs, while microRNA associated with tumor growth showed reduced activity. Genes known to be associated with anticancer activity also show increased expression in bats compared to other mammals.

# No stress aging

In the same way that the stress of flying may have helped drive the evolution of improved immune function, flight may also be part of the puzzle in understanding the unusual longevity of bats. As organisms grow older, their cells gradually become worn down or damaged by oxidative stress, leading to familiar age-related symptoms like skin wrinkles and cataracts. Some of this damage is caused by free radicals, naturally occurring atoms that have an unpaired electron in their outer shell, making them unstable and quick to react with other substances, including body cells. The most common group of free radicals are reactive oxygen species, which are produced as by-product during cell respiration. With extreme demands on respiration and metabolism, the act of flying should result in increased free radicals and cell damage. However, flying animals like birds and bats have evolved cellular adaptations to reduce and manage these free radicals, with a side benefit of reducing age-related cell damage. For example, little brown bats only produce about one-third the number of free radicals as the similar-sized, terrestrial white-footed mouse and short-tailed shrew.

Bats also have naturally high levels of antioxidants in their blood, special molecules that donate electrons to rampaging free radicals, thus neutralizing them and reducing cell damage. When researchers compared the amount of protein damage in cells of Mexican free-tailed bats, cave myotis, and mice, they discovered that bat proteins were overall more resistant to oxidative stress caused by free radicals.

## Sleeping for immortality

Hibernation may also play an important role in promoting longevity in mammals like bats. While bats generally live longer than most mammals of comparable size, there is still significant variation in maximum life span among bat species. When researchers examined different factors that might explain this variation, they found that body mass and hibernation were the best predictors of longevity. Specifically, bats that sequester themselves away and hibernate for a season live longer than bats that don't hibernate. In a follow-up study, the researchers looked at the changes in DNA expression caused by DNA methylation in big brown bats before and during hibernation. They found that genome regions associated with metabolism experienced decreased methylation during winter hibernation, such that the bats were essentially aging more slowly than when they were awake and active. This temporary slowdown in aging makes a difference: by hibernating for just one winter, big brown bats extend their epigenetic clock (and potential cell longevity) by almost 9 months.

Hibernation has similar effects in other hibernating mammals, with slowed epigenetic aging also demonstrated in hibernating yellow-bellied marmots. During hibernation, bats (and other animals) dramatically reduce their metabolic activity, giving them a chance to rest from the stress associated with flight. However, during their periodic arousals from torpor, bats may experience large

▲ The cellular proteins of cave myotis, like this cluster hibernating in a north Texas cave, are resistant to damage caused by free radicals in the body.

▲ By hibernating, this big brown bat is potentially extending its life span.

increases in oxidative stress, which they have adapted to handle. When awake, hibernating bat species have higher levels of antioxidant proteins than bats that don't hibernate.

## Long live the telomere

In addition to preventing cell damage from free radicals, another secret to bat longevity may be in the maintenance and repair of DNA. As we grow and age, our cells are continuously dividing. To protect the chromosomes during this process, we have evolved protective caps on the ends of the chromosomes made of repeating sections of nucleotides called telomeres. Telomeres act as buffers. When they get too short, it triggers a DNA damage signal that helps drive the aging process.

The telomeres of long-lived bats like *Myotis* don't experience the same age-related shortening as seen in most other mammals. This finding was surprising, as these bats also don't express the same telomerase genes as other animals. While telomerase is an important enzyme responsible for helping keep telomeres intact for as long as possible, it can also be associated with unregulated growth of cancer cells. Instead, bats like the greater mouse-eared bat and Bechstein's bat show upregulation of other telomere-maintaining genes like *ATM* and *SETX*, which help maintain chromosomes in the absence of telomerase.

## Oh Sugar, Sugar

Long doglike muzzles, short little ears, long goofy tongues, *and* important plant pollinators? Nectar-feeding bats sure check a lot of boxes when it comes to charismatic critters. But many nectar-feeding bats, especially those in the Neotropics, are unusual among mammals in that they are powered almost exclusively by high-sugar diets. If humans (and many other mammals) live on diets too high in sugar, we are at risk for developing undesirable health complications like type 2 diabetes. As far as we know, bats don't experience this same side effect from too much sugar.

Nectar-feeding bats have developed mechanisms for quickly converting the simple carbohydrates in nectar into energy. In one study, researchers discovered that Pallas's long-tongued bats begin converting sugar into energy within a few minutes of consumption. This is a significant adaptation, because in many animals, including humans, excess simple carbohydrates tend to get converted into more long-term energy storage like fat cells. Fruit-eating bats also rely on diets high in sugars like fructose. Using genomes from the Bat1K Project, researchers have identified two important genes in fruit- and nectar-feeding bats that may help them deal with their high-sugar diets. The first is a gene called *IMPA1*, which encodes an enzyme that, among other things, helps protects cells from stress associated with sugar metabolism. Several species of fruit-eating bats have duplicate copies of this gene, suggesting a link between this gene and high-sugar diets. The second gene with differences in bats is *SLC2A4*, which encodes a glucose transporter that helps move glucose from the bloodstream into muscle and fat tissue. Both Old World and New World fruit-eating bats show signs of positive selection on the part of the genome that encodes *SLC2A4*, contributing to the bats' improved ability to absorb glucose into the body and keep blood sugar levels down.

▲ A long-tongued bat slurps up some sugar water before being released.

# Learning to Speak

Humans' ability to learn and make new sounds as part of speech and communication is one of the things that sets us apart. The ability to both modify sounds and learn new sounds via imitation is rare in the animal kingdom, and much of what we know about this complex process has come from pioneering studies of birds (of which only some have this trait). While bird studies have provided many insights into the mechanisms and neurobiology behind speech and language, there are some important fundamental differences in the way birds' brains and vocal production work as compared to those of mammals. For example, instead of a larynx, birds produce sound from a comparable organ called the syrinx.

Bats, on the other hand, share laryngeal and communication-related structures with humans and other mammals. They also have sophisticated vocal systems that they use for both echolocation and communication, with the same or even more vocal flexibility than humans. As new genetic and neuroscience tools are developed, this positions bats as a useful model for understanding the evolution of language as well as speech-related disorders.

► Unlike in some other mammals, the pale speared-nosed bat can continue to learn and adjust its vocalizations into adulthood.

►► A multi-exposure shot of a Mexican funnel-eared bat's acrobatics in flight. Bats' flexibility and maneuverability in flight is the envy of humans trying to build better flying machines.

Direct evidence of vocal production learning has been documented in at least five families of bats. Pale spear-nosed bats display a trait called open-ended learning in which they can learn new vocalizations and vocal variations into adulthood. Other species, like the sac-winged bat, babble as pups, an important period in young bats and humans during which they practice vocalizations with feedback from adults. Advances in gene sequencing can provide new insights into genes associated with vocal production learning in bats. Researchers at the University of St. Andrews in the United Kingdom also recently developed methods for using magnetic resonance imaging (MRI) in bats, a tool that allows the observation of whole brain activity during vocalization in ways that are not possible using other neuroscience techniques. As a diverse group of animals, bats can help improve our understanding of language and communication traits across species as well as in humans.

# Bio-Inspired

The inspiration for Velcro, the sticky, interlocking loops and fasteners used for everything from hanging pictures on walls to fastening space suits, came about after the inventor observed the tacky, clinging nature of burr seeds. Observations of birds in flight helped drive the engineering of flying machines, starting with Leonardo da Vinci and eventually leading to the Wright Brothers. Throughout our history, human innovation has taken cues from the natural world, and it turns out bats make pretty good muses.

Birds may have been the main inspiration behind the jets and airplanes we see today, but bat flight offers some useful tricks too. With more than forty joints in their hand-wings and their thin, flexible, and strong wing membranes, bats are masters at quick, maneuverable flight. Trying to mimic this deceptively complex flapping flight, researchers at Caltech and the University of Illinois at Urbana-Champaign recently developed an autonomous Bat Bot. This bat-inspired drone tries to replicate the flexibility of bat wings with nine controlled joints, connected by a thin silicone-based simulation of bat skin. Researchers were able to program the Bat Bot to mimic bat maneuvers like straight flight and swooping-banking turns, though the bot can't hover (yet).

Likewise, researchers at the University of Cincinnati have been working to develop a drone that can use echolocation pulses to navigate. Currently, most drones must either be manually piloted using sight, video cameras, or satellites or by using machine learning algorithms. A downside of relying on video or satellites is when drones need to enter areas where these cues are not available, such as dark spaces and underground tunnels. Enter echolocation-inspiration!

But why mimic bat flight or echolocation in a drone? Drones with soft flapping wings—instead of the hard spinning propellers used by drone quadcopters—could be safer and quieter for use around people and wildlife. A similar idea applies for using sonar as a navigation tool, which might allow robots to scout potentially dangerous areas without putting humans at risk, like in collapsed buildings or underground surveillance.

Echolocation in bats is more varied and sophisticated than seen in most other animals, but echolocation is not unique to bats. Humans can also learn to use click echolocation to get around. With some training, both blind and sighted people are capable of learning how to echolocate. Although some have become highly adept at using sound to navigate, humans still have limitations for sound localization. For one, unlike bats, we can't swivel or move our ears around to better detect echoes from different angles, reducing our ability to discriminate finer details with sound.

Researchers from various universities and tech companies have tried to dream up assistive devices that could help humans better use echolocation for navigation. One device, developed by researchers from Stanford, MIT, and the University of California, Berkeley, consists of a wearable headset fitted with an ultrasonic emitter, a microphone, and two artificial ears, which looks a little

like a robotic bike helmet. The ultrasonic emitters create high-frequency pulses, and the microphones and associated software use time expansion to convert the sound into a frequency humans can easily hear. With limited training, people were able to make judgments about distance, laterality, and elevation. In another creative approach, computer science students at Wake Forest University developed a prototype of a watch that converts sound information into vibrations to alert the user to obstacles in the way. At this point, much of this technology is still a bit bulky or unwieldy, but researchers and engineers continue to dream of smaller or less obtrusive ways that sonar prosthetics might be used to help those who need them.

The natural world has long inspired humans in our inventions, engineering feats, and stories. By better understanding the complexities of the world of bats and how they came to be the way they are, researchers can use those lessons to help us face a changing and challenging world.

▲ Bats like this Mesoamerican mustached bat can use sound to detect movement and small objects in ways we humans can still only imagine!

# KEYSTONES

**Chapter 13**

The keystone is the wedge-shaped rock at the top of an arch that holds the rest of the structure in place. Without that stone, the arch would fall and crumble. Bats are categorized as keystone species, serving as insect predators, pollinators, and seed dispersers. From an ecological perspective, keystone species are those organisms that have disproportionately large effects on their environment relative to their abundance, and their removal results in drastic changes to that environment. While putting an exact value on bats is impossible, the environmental and ecological services they provide are not just critical to nature, but to humans as well.

▸ Voracious insect predators, species like the Mexican free-tailed bat are a farmer's best friend when it comes to pest control.

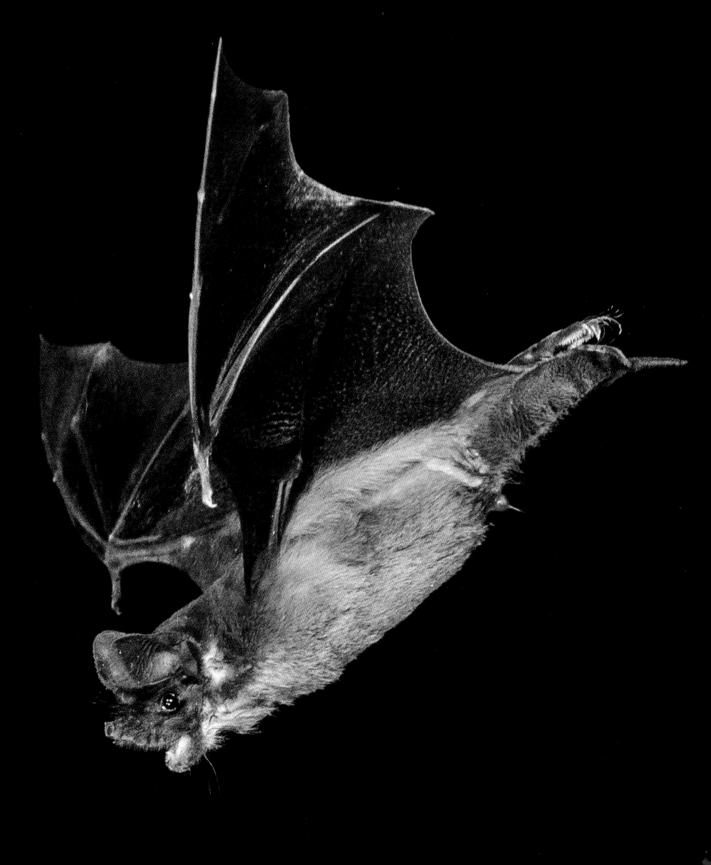

# 1-800-EAT-PESTS

Insect-eating bats have high energy demands and voracious appetites. That translates to a lot of insects eaten, including pests of both natural and agricultural systems. Some researchers estimate that bats can eat somewhere between 25 and 100 percent of their body weight in arthropods each night. For context, the current world champion hot dog eater, Joey Chestnut, has eaten 76 hot dogs in one sitting, the equivalent of about 16 pounds. If he was going up against a bat, he'd have to eat something like 150 pounds of hot dogs to keep up! Multiply that appetite by hundreds or even millions of individuals, and we start to see the pest control services bats provide.

Insect-eating bats feed on many species of agricultural pests, including certain moths, planthoppers, leafhoppers, June and click beetles, stink bugs, and weevils. In one study in the northeastern United States, researchers identified more than 600 insect species in the diets of little brown bats and big brown bats, 20 percent of which were known agricultural pests. This includes some insects that cause major damage to crops, such as the brown marmorated stink bug, western corn rootworm, and spotted cucumber beetle.

From their extremely large colonies, free-tailed bats worldwide consume pests of important agricultural crops like corn, cotton, and rice. Corn earworm moths (also known as cotton bollworm and tomato fruitworm) and beet armyworms make up large portions of Mexican free-tailed bat diets in the southern United States. Both insect species can cause significant damage to corn and cotton crops by feeding on leaves and corn ears in the larval stage. Similarly, wrinkle-lipped free-tailed bats in central Thailand consume both white-backed planthoppers and brown planthoppers, insects known to cause significant damage to rice crops.

While diet studies tell us the types of pests bats are eating, exclusion studies test whether and how bats affect insect populations. In these studies, researchers use netting to prevent animals like bats or birds from accessing a certain part of a crop or forest. After a set amount of time, they compare the number and types of insects present or the amount of plant damage between the netted and open plots. Exclusion experiments in crop systems give compelling evidence for bats as keystone predators. Excluding bats from plots of corn in southern Illinois resulted in 59 percent more corn earworm larvae per ear and

56 percent more damaged kernels per ear than paired plots that were open to bats. The researchers also found that by reducing insect damage to corn, bats also help reduce the growth of fungal pathogens, which take advantage of insect damage to invade the plants.

When bats were excluded from areas of a coffee plantation in Mexico, total arthropod densities increased by 84 percent per coffee plant in the wet season. Another study in coffee plantations on the slopes of Mt. Kilimanjaro in Tanzania revealed that when both bats and birds were excluded from plots, there was a 9 percent reduction in fruit set, though this study did not separate the effects of bats and birds. Even more dramatically, exclusion of bats and birds from cacao plants in Sulawesi resulted in a 31 percent decrease in cacao yield.

▲ Big brown bats eat many insects that are garden and crop pests.

Exclosure experiments have also demonstrated that bats are effective at suppressing insect damage in rice crops in India and reducing leaf and grape damage in vineyards in Chile. So, next time you enjoy a hot cup of coffee, relax with a glass of wine, or snack on a chocolatey treat, don't forget to give a little thanks to the bats.

## Show me the money

By reducing plant and crop damage, bats provide large economic benefits for farmers and other people across the world. In rice paddies in the Mediterranean, soprano pipistrelles like to munch on the rice borer moth, one of the major rice pests around the world. Following the installation and occupation of bat boxes in the eastern Ebro Delta of northeastern Iberia in Spain, the rice borer moth population in the area was significantly reduced; using the market values of rice crops and pesticides, the researchers estimated the value of bat activity at around 52 Euros (or about $57) per acre in reduced pesticide use. Likewise, Australian researchers estimated that populations of Gould's wattled bat consume between 77 and 119 tons of moths a year, saving cotton farmers about $63 million annually from damage. Similarly, bat predation on stink bugs in South African macadamia orchards—the world's largest producer of macadamia nuts—may help farmers save over $1500 per acre per year.

Even small reductions in pest damage can yield important economic benefits to farmers. In an exclusion study conducted in central Chilean vineyards, grape damage was 7 percent lower in control plots where bats had access to insects, yielding an average economic benefit of over $600 per acre per year. In another estimate of bat pest suppression, even if each night a single bat in Brazil only ate two fall armyworm moths (a frequent pest of corn, sorghum, soybeans, and sugar cane), they could reduce the moth population by 20 percent. Considering that these moths can damage between 20 and 100 percent of crops, bats could save Brazilian farmers about $390 million per harvest.

Wrinkled-lipped bats in Thailand like to feed on planthoppers, a major pest of rice crops in the region. With an estimated population size of about 8 million, the bats may prevent an average loss of 2892 tons of rice per year (valued at over $1 million) and protect food for more than 36,000 people per year. In the United States, Mexican free-tailed bats are the stars. In the cotton fields of south-central

Texas, researchers estimated the annual savings of avoided cotton damage to be between $500,000 and $700,000. This includes both estimates of crop damage and the reduction of pesticide that would be needed to control the same number of insects predated by bats. Extrapolating these values from just south-central Texas to croplands across the continental United States (while accounting for regional differences in bat activity and crop types), it's estimated that the economic value of bats to the nation's agricultural industry is between $3.7 and $53 *billion* a year, with a mean estimate of about $22.9 billion a year.

## Beyond agriculture: fields and forests

Bats don't reserve their pest control services for agriculture. They are also important for pest suppression in forests. Using netting to exclude either birds or bats from plots of forest in Panama, Margareta Kalka and colleagues found that plots excluding bats contained as much as 153 percent more arthropods than paired control plots, and netted plots lost about 13 percent of their leaf area. Diet analysis confirms that tropical bats eat leaf-damaging insects like caterpillars, crickets, katydids, and scarab beetles. The effects of bat (and bird) predation on insects in tropical forests could be especially important for tropical forest restoration, where leaf damage hurts the growth and survival of young and newly planted trees.

◄ A big-eared bat (*Micronycteris* spp.) roosts in a culvert in Panama. These gleaning bats can detect caterpillars on trees using echolocation and help protect trees from leaf damage.

Bats also benefit forests in temperate regions of the world. In one of the few exclusion experiments to test the effects of bat insect suppression in temperate forests, Elizabeth Beilke found that when bats were excluded from forest plots in south-central Indiana, insect densities were three times greater than in control plots. Even more significantly, bat exclusion resulted in five times more seedling leaf damage compared to the controls. Diet studies of forest-dwelling bats also confirm their insect control. For example, in beech forests in central Italy, barbastelle bats and long-eared bats were found to eat a range of insects that damage trees. These studies highlight the role that insect-eating bats play in maintaining healthy forests.

## Peter, Peter, mosquito eater

Collectively, bats certainly eat a lot of insects, including small flying insects like mosquitos. But are bats champions of mosquito control? In a paper published in 1960, Don Griffin and colleagues were testing his hypotheses about how bats navigate and capture prey using echolocation. They released bats and either mosquitos or fruit flies into a flight cage in the laboratory and then observed how bats pursued and captured the insects. By monitoring the bats in these enclosed areas, the researchers estimated that bats could catch mosquitoes at rates ranging from about 2 to 10 mosquitoes per minute. If we assume the maximum, then bats could potentially eat as many as 600 mosquitoes in a single hour. In another observation in a mosquito swarm in northern Sweden, one researcher recorded bats catching 20 insects per minute. It's *this* value that has been extrapolated to the popular statistic of bats eating 1000 mosquitoes an hour. Further extrapolation can lead to claims like 12,000 mosquitoes per night or 4.3 million per year. Amazing, right?

Maybe not. These estimates have a couple flaws. First, in the Griffin study, the extrapolated values were based on maximum reported intake, not the average. Let's go back to the world champion in hot dog eating, Joey Chestnut. His world record is 76 hot dogs in 10 minutes, set in 2021. Based on this rate, Chestnut should be able to eat 456 hot dogs in an hour! That's impossible. So, while bats can eat a lot of insects at once, it is unlikely that they can maintain these high consumptions rates over a long period of time. Second, most of us eaters probably can't get anywhere close to that 10-minute rate of 76 hot dogs. Estimating

◄ Little brown bats might not actually eat 1000 mosquitoes an hour, but they do feed on a range of pesky insects.

total effect from one high-performing individual skews estimates of everyday behavior. Finally, the Griffin study was done by releasing hungry bats into a room filled with mosquitoes. In the wild, bats are eating a range of different insects and other arthropods, depending on their body size, morphology, flight abilities, foraging habitat, and time of year. This is important because mosquitos are small and not particularly high in energy content. While some small insect-eating bats might be able to meet their daily energy needs by consuming a diet of mostly mosquitoes, it's unlikely larger bats can do the same.

Bats might not necessarily be mosquito-eating fiends, but that doesn't mean they don't contribute to mosquito control. Diet studies that looked for insect parts in bat poop generally found that mosquitoes made up less than 10 percent of a bat's overall diet, but mosquitoes are difficult to identify in poop. A 2018 study used molecular techniques to detect insect DNA in bat poop, reporting that mosquitoes—representing up to 15 different species—were present in the poop of about 72 percent of little brown bats sampled in Wisconsin. Other studies have found evidence of mosquitoes in the diets of Seminole bats, evening bats, and southeastern myotis, including six mosquito species associated with diseases like West Nile virus, dengue, and Zika virus.

So far, only one study has tested the effect of bat predation on mosquito life-style. In a field experiment, researchers released northern long-eared bats in mesh enclosures that contained the bats but allowed in mosquitoes and other insects. They also added plastic containers filled with water and hay, the type of habitat that mosquitoes use to breed. After allowing the bats to live and forage in the enclosures for 9 nights, they counted the number of eggs laid by mosquitoes in the plastic containers, finding a 32 percent decrease in enclosures that contained bats. While compelling, this work was done only in one species of bat and at a very limited spatial scale. Further experimental and field studies are needed to determine if the prevalence and abundance of mosquitoes in bat diets are enough to significantly impact mosquito populations.

# Guano Happens

In cave ecosystems, where levels of life-sustaining nutrients can be difficult to come by, bats are keystone species as providers of guano. That's right, bat poop serves as an important food resource for cave-dwelling organisms, especially in dry caves that don't receive outside nutrients via streams or rivers. A range of invertebrate detritivores (animals that feed on dead organic material), including mites, beetles, springtails, isopods, moth larvae, and flies, feed on guano. But invertebrates aren't the only critters that will feed on bat guano. The western grotto salamander that lives in cave habitats in the central and southeastern United States may regularly feed on bat guano in their larval stage. Turns out, bat guano is nutritionally comparable to invertebrate prey—and a lot easier to capture.

► The western grotto salamander is one of the many cave-dwelling animals that consume bat guano.

With its high nitrogen and phosphorus levels, the poop of insect-eating bats can also serve as a nutrient source for plants. Bats help redistribute these nutrients over the landscape via the so-called pepper-shaker effect, sprinkling poo over forests and fields as they fly about at night. Bat guano can also act as an organic fertilizer for natural vegetation and crop plants. In a greenhouse experiment, moderate applications of guano from Mexican free-tailed bats promoted growth in some native Texas grass species, like Indian grass, but either reduced growth or had no effect on others, like little bluestem or prairie coneflowers. Bat guano–based fertilizer is also used by farmers to increase growth in papaya, eggplant, and moringa in Cambodia and in tomato plants in Niger. While bat guano can be beneficial to human crop systems, the widescale application as fertilizer needs to be approached carefully. Unsustainable harvesting can disrupt potentially fragile cave ecosystems that rely on guano, disturb bat colonies, and expose workers to potentially unsafe working conditions.

## Pooping far and wide

In sprinkling their seed-filled guano over the landscape, fruit-eating bats in tropical and subtropical regions act as seed dispersers. Many fruit-producing plants rely on vertebrates to help distribute seeds away from the parent plant, where seedlings would need to compete against nearby mature plants. In removing fruits and carrying them away from the parent tree, bats can help move seeds large distances. In Panama, for example, Jamaican fruit-eating bats carry fig fruits up to 250 m from the tree to feed. Some Old World fruit bats can travel much farther distances, even moving seeds across the ocean from one island to another.

Unlike birds, who usually poop while perching, bats poop and spit out seeds while in flight, increasing the chance that seeds will be moved to a new place away from the parent tree. Bats also fly across disturbed or clear-cut areas, sprinkling seed-filled guano as they go. In the Neotropics, species like Seba's short-tailed fruit bat and the little yellow-shouldered bat help transport the seeds of early successional plants like *Piper, Solanum*, and *Cecropia*. These fast-growing plants are examples of pioneer species that initially colonize disturbed areas, setting the stage for later forest growth. In proof that bats can disperse seeds to favorable environments, forest plots in Costa Rica near the roosts of short-tailed fruit bats had both higher species diversity of *Piper* plants and lower total herbivore leaf damage than plots without bat roosts.

Old World fruit bats don't disperse as many early successional plant species, but they do disperse the seeds of larger plants found in more mature forests, such as figs. In an old-growth lowland rainforest in Malaysia, an estimated 13 percent of tree species are at least partially dependent on bats for seed dispersal. Fruit bats can also move seeds long distances. When straw-colored fruit bats are undertaking their long migration across central Africa, they can move as far as 88 km in a day. Potential seed dispersal distances—estimated based on bat movement and the average time it takes a seed to move through the digestive system of a bat—could be greater than 75 km, a distance roughly four times farther than maximum distances estimated for similar-sized frugivores that are not bats. During the dry season, when straw-colored fruit bat colonies can reach over 150,000 individuals, each bat can produce a seed rain of 26.2 dispersal events (a fancy way of saying poops) per kilometer in a single night.

Estimating the economic value of seed dispersal services by bats is challenging, especially since seed germination and plant growth is dependent on more than just how far or where a seed gets dispersed. However, across the world there are many human-cultivated and otherwise valued species that are

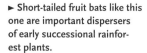

▶ Short-tailed fruit bats like this one are important dispersers of early successional rainforest plants.

dispersed by bats, including fruit crops like papaya, cashew, soursop, and dates. Bats also disperse the seeds of the tropical almond tree, which grows through-out India and Southeast Asia and is used as a source of timber, edible seeds, tan-nins for dye, and bark and seed extracts for medicine.

# Face Full of Pollen

Pollination occurs when pollen is transferred from the anther (male part of a flower) to the stigma (female part of the flower) to fertilize the plant. Some plants can self-fertilize, with the pollen from one flower fertilizing either the same flower or another flower on the same plant. Other plants need their pollen to be moved to the flower of another plant to reproduce, and still others can do a combination of both. Pollinators like bees, butterflies, birds, and bats assist with this process by moving pollen that gets stuck to their bodies while feeding to a new flower or plant. Researchers estimate that bats provide ecosystem services to over 500 species of plants worldwide as seed dispersers and pollinators, with some bats serving as both. Nectar-feeding bats in the families Phyllostomidae and Pteropodidae are the main bat pollinators throughout the world.

What makes bats such good pollinators? Compared to insects, bats are large, which means there's a lot of surface area for pollen grains to stick to as they visit flowers. Bats can also move pollen grains long distances, helping to ensure gene flow between forest patches and even islands. Having a pollinator with a strong preference for a certain plant can also be important for pollination success, as specialist pollinators minimize the risk of the stigma getting clogged with pol-len from the wrong species. Although some bats show strong preferences for certain plants, even those with preferences take advantage of various plants that flower at different times of the year.

Nectar-feeding bats aren't the only ones who visit flowers. Although most of their diet consists of scorpions and other arthropods, pallid bats also feed on the fruit and nectar of cactus flowers in some parts of their range. The cardon cactus has large, white, slightly stinky flowers that are pollinated by bats, mostly long-nosed bats. When nectar-feeding bats feed from these cactus flowers, they gracefully hover above the blossom, slurping nectar and incidentally collect-ing pollen. When pallid bats do it, however, they end up bear-hugging the base of the flower while they stick their whole face into the flower opening. This

► The head of this long-tongued bat in Belize is covered in flower pollen.

◄ While feeding on a flower, this grey-headed flying fox has gotten a face full of pollen that it will transfer to other flowers.

◄ A pallid bat with a face full of cactus pollen is removed from a mist net.

less-than-graceful method results in significantly more pollen grains sticking to the bat and thus getting transferred to the next flower.

Bats are also critical pollinators of mangrove forests on the islands and coastal areas of Southeast Asia. Consisting of trees and plants that are adapted to wet soils with distinctive long, spindly roots, mangrove forests are hugely important coastal ecosystems. Not only do they provide habitat for a range of terrestrial and aquatic plants and animals, mangroves help protect shorelines from potentially damaging storms, winds, and floods by preventing erosion and trapping sediments. Mangrove apple, one of the most widespread mangrove species in the Indo–West Pacific region, is mainly pollinated by bats. Although mangrove apples are also pollinated by other animals like moths, one study reported that pollination by bats resulted in the highest fruit set and seed production.

Some agricultural plants are also pollinated by bats. For example, in Southeast Asia bats are the main pollinators of the durian tree. Also called the king of fruit, durian is a large, thorny fruit with a distinctly pungent odor, creamy texture, and unusual flavor. Common bat visitors include the cave nectar bat, island flying fox, black flying fox, and northern blossom bat. A few other species of nectar-feeding bats and flying foxes are also observed to feed on durian flowers, depending on location and time of year.

▼ The white flowers of the durian tree are gently pollinated by flying foxes and other bats.

On Tioman Island in Peninsular Malaysia, the endangered island flying fox was the primary pollinator of both cultivated and wild durian trees, resulting in increased fruit set. While durian farmers worry that bats will damage the plant's flowers, infrared video recordings from this study found that, despite their size, island flying foxes were largely nondestructive and surprisingly delicate, feeding only on the nectar and rarely damaging the actual flower. An Indonesian study on durian-bat interactions led by Sheherazade (Shera, a Sulawesi student who is now a PhD student and Fulbright scholar at University of California, Berkeley) found similar results. When Shera and her team restricted bat access to flowers by blocking them with netting, fruit set decreased compared to unobstructed flowers. From this study, they also estimated the economic value of bat pollinators to be about $289 per acre of durian, or about $450,000 a year.

Other economically and culturally important trees throughout the Paleotropics are known to rely at least partially on bats as pollinators. The African sausage tree grows in open woodlands and along waterways in southern Africa. The species is popular as a shade tree along streets in South Africa, and the fruit

▲ Wahlberg's epauletted fruit bat is an important pollinator and seed disperser of African plants.

and leaves are used for a wide variety of cultural and medicinal purposes including traditional African beer, as a facial cosmetic, and to treat skin and stomach ailments. The African sausage tree produces dark red flowers that are visited by bats like epauletted fruit bats and birds like sunbirds and bulbuls. In one study in South Africa, bats were found to deposit twenty times more pollen per flower visit than birds. Bats also help pollinate the iconic baobab tree, an economically and ecologically important tree in African savannah ecosystems. Bat visitation to baobabs in Benin increased fruit set, whereas baobabs in other areas may be primarily pollinated by hawkmoths and bees.

In arid subtropical and tropical regions of North and South America, bats are the primary pollinators of many economically and ecologically valuable plants in the agave and cactus families. In Mexico, these plants are pollinated by nectar-feeding phyllostomid bats, including the long-nosed and long-tongued bats. Most famously, these bats are the main pollinators for *Agave tequilana*, the source of commercial tequila. In Mexico and Central America, more than sixty species of agave plants are used to make alcoholic beverages such as pulque, mezcal, and bacanora, and the plants' fibers and stalks are used for weaving and building fences. Pollination by bats also has been shown to enhance both the quality and yield of cactus fruits, notably pitaya (dragon fruit). When bats were excluded from pitaya flowers, fruit yield decreased by 35 percent and the fruits were 46 percent lighter and 14 percent less sweet than when pollinated by bats. Since both yield and fruit size are important for setting price in the pitaya market, pollination services by bats in this system are estimated to be worth as much as $6175 an acre.

## Banana myths

Globally, people consume more than 100 billion bananas a year, with Americans eating an average of 27 pounds of bananas per person per year. Although native to Southeast Asia, bananas are now grown in tropical regions around the world. Wild banana species are pollinated by a range of vertebrates, including nectar- and fruit-eating bats in both their native range and in introduced areas like Latin America. However, the bananas that might be sitting on your countertop or that you see in the supermarket were probably *not* produced with the help of a bat. That's because most of the bananas exported to countries like the United States

are Cavendish bananas, which have been bred to not have seeds and only reproduce via cloning. This asexual reproduction means that, though cultivated plants do bear nectar-producing flowers, those flowers don't produce pollen. Bats might not have helped pollinate your bananas, but they do still pollinate wild bananas around the world, including local varieties in Africa and Asia that are consumed domestically instead of being exported.

In Neotropical countries like Nicaragua and Costa Rica, some nectar-feeding bats regularly drink from cultivated banana flowers. Banana plantations might seem like a great place for bats to forage, but they can be the bat equivalent of a fast-food restaurant. The food is cheap and easy to access, and bats that forage mostly in conventional banana plantations are heavier than those foraging in wild forests or organic banana plantations. However, those same bats that mostly fed from banana plantations also had modified gut microbiomes and a diet that may lack protein, since farmed banana flowers don't contain protein-rich pollen.

▶ A long-tongued bat feeds from a banana flower in Belize. Bananas are not native to the Neotropics, though some species of nectar-feeding bats have learned to feed from their flowers.

▲ An Indian flying fox visits a banana flower.

# CONSERVE

**Chapter 14**

Our knowledge of the ecosystem services provided by bats, including pollination, seed dispersal, and insect control, is far from comprehensive. There's still a lot to learn about their role as insect suppressors in natural ecosystems like forests, as well as the effects of bat seed dispersal and pollination on ecologically and economically important plants. Bats are important to people, helping provide us with food, medicine, and timber. However, habitat loss and fragmentation due to habitat loss can impact bats. So as much as we need bats, bats also need us.

▶ A critically endangered Jamaican flower bat leaves its cave.

We are in the middle of a global biodiversity crisis, a consequence of human activities including land-use change, overexploitation, introduction of invasive and domesticated species, and climate change. Like many organisms across the globe, bats are also facing conservation challenges from these threats. Bats' high diversity and global distribution, combined with their elusive habits and a lack of bat research in many regions of the world, make identifying and addressing these threats a huge challenge.

A total of 1331 bat species are currently included on the International Union for Conservation of Nature (IUCN) Red List. Of these, 221 (17 percent) are considered threatened. The IUCN Red List is an inventory of the global conservation status of species, in which conservation science experts categorize the extinction risks based on existing data, population estimates, and potential threats. Species are considered threatened with extinction if they fall into one of three categories: Critically Endangered, Endangered, or Vulnerable. Aside from the threatened species, another 235 bat species (18 percent) are currently listed as Data Deficient, meaning there is not enough information about the distribution or population status to make an estimate of the species' extinction risk. Independent of extinction risk, the population trends of 52 percent of the bat species on the IUCN Red List are unknown, whereas about 24 percent are thought to be decreasing, including 775 species that are currently classified as Least Concern.

These gaps in knowledge can have potentially huge consequences for bat conservation. In a recent study, researchers used computer models to predict which bat species might be at the highest risk of extinction. Of the ten species they identified as high risk, six are currently categorized as Data Deficient. The researchers also estimated that at least 10 percent of the bat species designated as Data Deficient could be threatened, highlighting the difficulty in identifying and addressing bat conservation issues.

# Invasive Predators

As a result of both intentional and unintentional actions, humans have rearranged ecosystems by mixing together species that did not previously interact. Although some of these introduced species can be relatively harmless, many become invasive, spreading in a way that damages the existing ecosystem. Introduced predators are a major concern for bats. Compared to other mammals, bats tend to experience relatively low predation rates from native predators like owls, hawks, and snakes, due to their nocturnal habits and ability to roost in hard-to-access spaces like caves.

One of the most widespread and worst invasive predators is the domestic cat. Free-ranging pets and feral cats are a global threat to biodiversity, with estimates suggesting they have contributed to about one-quarter of vertebrate extinctions since the year 1500, mostly of birds and small mammals. Quantifying the effect of cats on bat populations is challenging, but there is little denying that cats will attack and kill bats, with at least eighty-six species identified as preyed upon or threatened by cats. Just a single cat can have potentially devastating consequences on a bat colony. In an extreme example, a single male housecat was estimated to have killed at least 102 threatened New Zealand lesser short-tailed bats over the course of 7 days before being captured. Bats that inhabit islands are particularly vulnerable to invasive predators like cats, due to their restricted geographic ranges and naivety to introduced predators. Cats have been captured on game cameras catching bats at a protected cave in Jamaica, the only known roost for the Critically Endangered Jamaican funnel-eared bat. Surprisingly, cats will also prey on large fruit bats like the Christmas Island flying fox and the Ryukyu flying fox, both categorized as Vulnerable by the IUCN.

▲ To protect bats and other wildlife, keep cats indoors and only allow them outside on a harness or under close supervision.

# Fungal Invasion

In March 2007, biologists from the New York Department of Environmental Conservation drove out to a hibernaculum about 20 miles from the town of Howes Cave, New York, to conduct a winter bat survey. What they found were thousands of dead bats on the floor of the cave, some with a white, powdery substance on their noses and wings. Dead bats were also reported at three other caves in the area, which quickly raised alarms. Named white-nose syndrome for the white growth observed on the noses of dead bats, this mystery disease spread to 8 more states within just 2 years. Close study of tissues and anatomy from dead bats revealed that the white power was a new species of fungus growing on the skin of bats. By the time the first peer-reviewed article

▲ This hibernating tricolored bat is infected with white-nose syndrome.

describing this bat-killing fungus was published in 2008, bat populations in Connecticut, Massachusetts, New York, and Vermont had declined by at least 75 percent.

In the 17 years since white-nose syndrome was first observed in that New York cave, the fungus has spread to thirty-nine states and eight Canadian provinces and is known to infect at least twelve species of hibernating bats. Now known as *Pseudogymnoascus destructans*, the deadly fungus was unknown to science until it started causing problems for North American bats, though it has since been identified in samples from Europe and Asia, including from a 100+ year old bat specimen that was collected in France in 1918. How it got to North America is still unclear, though it was most likely mediated by humans either through accidental transport of an infected bat or transfer of fungal spores on caving gear or equipment.

Unlike many other fungi that survive by breaking down dead and decomposing matter, *P. destructans* grows on living tissue. Temperate bats are exposed to the fungus in late fall and early winter when they return to their hibernacula from their summer habitat. *P. destructans* can only grow at temperatures below 20°C (68°F), so it doesn't infect active, flying bats. It's not until bats settle into hibernation and drop their body temperatures that the fungus can strike. As bats are hibernating, any *P. destructans* spores on the bat's fur and skin begin to grow, eventually resulting in the diagnostic white fuzz on the noses and wings of the infected bat. However, it's not the fungal growth that is directly responsible for killing the bat. Instead, *P. destructans* infection triggers a cascade of physiological effects that result in more frequent arousals, loss of fat stores, evaporative water loss, and, ultimately, starvation. In some bats, an immune overreaction may also lead to additional tissue damage. For bats that manage to survive the winter with the fungus, fungal loads quickly drop following emergence from hibernation (within 10 days), becoming undetectable by midsummer. Unfortunately, even after bats leave the hibernacula for their summer grounds, *P. destructans* continues to persist in the cool cave or mine environments, where it can reinfect bats the following fall.

White-nose syndrome has had devasting impacts on bat populations across North America. Prior to the arrival of *P. destructans* in New York, little brown bats were the most abundant and widespread bat in the northeastern United States. Since 2007, little brown bat populations in the eastern United States have

declined by more than 90 percent, as have populations of the northern long-eared bat and tricolored bat. Northern long-eared bats, which were first listed as Threatened under the Endangered Species Act in 2015, were recently up-listed to Endangered due to these severe declines. The US Fish and Wildlife Service also recently proposed listing the tricolored bat as Threatened. Other species have also been affected, with big brown bats experiencing declines of around 35 percent due to white-nose syndrome and the endangered Indiana bat experiencing variable declines of between 28 and 84 percent across their range.

▼ Samples taken from the wings of bats and cave surface during winter surveys are sent to the laboratory to test for the presence of *Pseudogymnoascus destructans*.

▲ Indiana bats were already listed as Endangered under the Endangered Species Act when white-nose syndrome was discovered. Although some populations experienced greater declines than others, Indiana bats have been less affected than some similar species.

## Cause for hope

It's early September 2021, and I am on the Keweenaw Peninsula, the northernmost part of Michigan's Upper Peninsula. Projecting like a little thumb into Lake Superior, the area is dotted with caves and old copper mines used by hibernating little brown bats, big brown bats, northern long-eared bats, and tricolored bats. Since *P. destructans* was detected in the region in the winter of 2013–2014, northern long-eared bats and tricolored bats have all but disappeared

from the area and the huge colonies of little brown bats have dwindled to a fraction of what they once were. But as I stand outside a mine entrance at dusk and hear the swooshing of wings, I can't help but feel a bit of hope.

Some bat species don't appear to be susceptible to white-nose syndrome at all, whereas others have experienced severe declines. Scientists are still trying to understand what might drive this variation. Species preferences for different temperatures and humidity during hibernation may be part of the puzzle, as *P. destructans* grows best in warmer hibernation temperatures (around 16°C, 61°F) and higher humidity. Variation in hibernating patterns, such as how often bats naturally arouse from torpor, may also affect species susceptibility. The *P. destructans*–resistant eastern small-footed bat and endangered grey bat enter torpor for short periods and don't drop their body temperatures as low as other species, making it more difficult for the fungus to take hold.

Certain bacteria that naturally grow on the skin of more resistant bat species might also slow the growth of *P. destructans*. Researchers have tested the possibility of harnessing this bacterial power as a probiotic treatment to reduce impacts of white-nose syndrome on declining bat populations. Both laboratory and field studies have shown promise, with bats treated with a probiotic spray showing reduced disease severity and increased survival. Other treatment strategies being tested include naturally occurring antifungal volatile organic compounds, vaccines to boost bats' immune responses against the fungus, and artificially cooling alternate cave sites to give bats new places to hibernate. While each of these approaches show promise, applying them at a large scale is still challenging.

Although little brown bat populations remain much smaller than they were, there is some evidence that the populations may be stabilizing. One study found that the intensity of infection was lower in populations where *P. destructans* had been around for a while compared to recently invaded and declining populations, a sign that adaptation may be possible. Other studies have searched the genome of recovering bats to see if surviving populations might develop resistance to this invasive pathogen. Small changes have been detected in the genomes of some populations, mainly in genes associated with immunity, hibernation pattern, and fat use, although no two populations appear to be adapting in the same way. Even if bat species are beginning to adapt to this disease pressure, evolution acts on populations, not individuals. Bats reproduce slowly, meaning any advantageous, heritable changes are going to be slow to move through the population.

## Fat bats

In 2019, Dr. Tina Cheng, a researcher at Bat Conservation International, led a study investigating bat fat. Cheng and her collaborators observed that bats with higher amounts of body fat were more likely to survive the winter, even with white-nose syndrome. This gave the conservation scientists an idea: what if we could help the bats at high risk of *P. destructans* infection get fatter? In addition to mating during prehibernation swarming, bats spend a lot of time eating to increase their body mass and fat stores to help get them through the winter. And, thus, the Fat Bat Project was born. Using ultraviolet lights known to attract insects, the idea was to create experimental prey patches near bat hibernacula and make it easier for hunting bats to find food. Initial trials of placing lights near known hibernacula showed promise, with bat foraging activity between three and eight times higher at ultraviolet light lures compared to unlit controls. While promising, it remains to be tested if creating these high-value bug buffets will translate into fatter bats or increased overwinter survival.

# Helping WNS-affected bats

Unfortunately, it is highly unlikely that using a vaccine or probiotic or putting out an ultraviolet light is going to be a grand solution. So, what can the average person do to help bats affected by white-nose syndrome? One way is to provide quality foraging and roosting habitats for bats. This can range from helping with habitat restoration efforts in your area to promoting native plant growth and reducing pesticide use in your backyard.

It also means doing our part to minimize the continued spread of *P. destructans* to naïve bat habitat by properly cleaning and decontaminating clothing, boots, and other equipment used while caving. A short precleaning to remove mud and sediment followed by application of commercially available disinfectants is very effective in clearing *P. destructans* from gear. Cavers and tourists should consult current versions of the National WNS Decontamination Protocol published by the White-Nose Syndrome Disease Management Working Group (WhiteNoseSyndrome.org). Finally, avoiding unnecessary disturbance of bats at hibernacula is also important to reduce the costs of them waking up. Always respect signage or gates blocking entry at caves or mines.

# Protect the Bat Cave!

Half the world's bat species rely on caves, mines, or other underground habitat for at least part of their life, and about 15 percent of these species are currently classified as Threatened by the IUCN. The biggest threat to caves and their bat inhabitants is disturbance. This includes intentional vandalism and destruction, as well as extractive practices like limestone quarring, guano harvesting, and cave tourism. Noise and human activity during hibernation can be especially problematic, because causing bats to wake up frequently from torpor depletes those precious fat reserves. Disturbance can also lead to roost abandonment, forcing bats to seek out alternative and often inferior roosting locations. Human activity and modifications associated with tourism, such as the installation of lighting or walkways, can also change the microclimate of cave areas by altering light levels, airflow, and temperature or humidity, making the space less hospitable for bats. Nearly the entire population of the endangered Fijian free-tailed bat is limited to only one known roost on the island of Vanuatu, and the critically endangered Jamaican flower bat is known from only two caves in Jamaica, meaning that even small disturbances at these sites have the potential to impact the species as a whole.

Protecting caves is challenging. In Europe and North America, academic societies and other nongovernment organizations, like the National Speleological Society in the United States, have developed guidelines to minimize the impact of cave exploration and recreation on cave ecosystems. Bat-focused conservation programs like the Southeast Asian Bat Conservation Research Unit and Latin American Network for Bat Conservation are also working to increase public education, cave protection, and partnerships with industry to reduce the impacts to critical cave habitat. Installing gates and fences at mine and cave entrances is another common measure used to both protect cave habitat and public safety. Most of the research on the effectiveness and impacts of these gates on bats has been limited to North America and Europe, with future research needed to assess the pros and cons of installing gates in other areas of the world.

▲ This Buddhist shrine built inside a cave in northern Thailand is also a tourist destination. The larger caverns at this site are home to at least two species of horseshoe bats.

# Bats and Forests

Loss of forest habitat is a severe threat to global biodiversity, including bats. In most places, deforestation presents as habitat fragmentation, creating a mosaic of primary and secondary forests, edge habitats, and agricultural and pastoral lands. Among the well-studied New World leaf-nosed bats, gleaning insectivorous and carnivorous species are the most negatively affected by forest fragmentation, whereas fruit- and nectar-feeding bats that forage on cultivated plants are less impacted or even benefit from habitat fragmentation. Similar patterns are also seen in other tropical regions. In Kenya, for example, insect-eating bats that prefer closed canopy habitats avoided smaller forest fragments, whereas fruit-eating bats were less picky.

In temperate forests, clear-cuts for timber harvest in which all trees from an area are removed can temporarily lead to increased activity of bat species that prefer to forage in more open spaces, but it negatively affects interior forest specialists. Selective logging practices, in which only certain trees are removed, can help preserve some of the diversity of old-growth forests for many taxa. This logging still creates potentially fragmented landscapes, which studies suggest can be both good and bad for New World bat species. Few studies have examined

▲ A patch of forest has been cleared near a village at the edge of the Calakmul Biosphere Reserve in Mexico. Some species of bats are more tolerant of this type of edge habitat than others.

◄ Vast tracts of land are being cleared for crops and livestock grazing in Belize.

the effect of logging on African bats, though one study in Uganda found no significant difference in the activity of the insect-eating banana serotine between primary and logged forests. Mixed responses to logging were also observed in Southeast Asia, though unlogged forests generally had a higher diversity and abundance of forest-dwelling bats. Advances in capture techniques and acoustic monitoring in these regions can help improve our understanding of how tree and plant harvesting impacts bats.

# Crop Fields Everywhere

Land conversion for agriculture is a major contributor of habitat loss and modification, with nearly 40 percent of the Earth's terrestrial land cover estimated to be in some type of agricultural production. This includes not just field crops like corn or cotton, but also agroforestry products (coffee, cacao, and tea), orchards and forest plantations, and pastoral lands for grazing cattle and other livestock. As pollinators and pest control agents, bats are hugely important to human agriculture. However, agriculture is considered a major threat for more than 50 percent of threatened bat species.

In traditional agroforestry systems, such as coffee and cacao, cultivated plants are grown among native shade trees. This makes them potentially compatible with conservation efforts, as they often still resemble natural forest habitat in some respects—such as canopy cover or tree density. There's evidence that at least some bats benefit from shaded agroforestry, mostly fruit- and nectar-feeding New World leaf-nosed bats. In contrast, croplands dominated by one plant species or wide expanses of pasture without trees or other structural variation tend to have more negative effects on bat diversity and abundance. Some of these effects can be mitigated by landscape elements such as hedgerows, which provide protected corridors for bat movement. Studies conducted in Germany and South Africa have also shown that restoring or constructing wetland habitat in agricultural landscapes can help support bat populations.

## Tequila and bats: a partnership

Long-nosed bats rely on blooming columnar cacti and agave plants to fuel their long seasonal migrations each year. However, the species' critical desert habitat is under threat from grazing and agricultural development, including small- and large-scale production of tequila and mezcal. And, while it seems like vast fields of agave plants would be good for these nectar-feeding pollinators, unfortunately that is not the case. In cultivated agave, farmers clip the flower stalks

▼ *Agave tequilana* grows in a field in Jalisco, Mexico.

to prevent sexual reproduction in the plant, instead using clonal reproduction to propagate plants. This results in a food desert for bats, with large areas of no flowering plants.

Reliance on clonal reproduction reduces the genetic diversity of cultivated agave, making it susceptible to bacterial and fungal diseases, insect pests, and environmental variation. Recognizing the critical link between bats, agave, and people, bat researchers from the National Autonomous University of Mexico joined forces with the Tequila Interchange Project, a nonprofit focused on the preservation of traditional tequila and mezcal practices. Letting just 5 to 10 percent of agave plants flowers helps provide critical foraging habitat for local bats and promote genetic variation within the crops. Tequila made from agave plants grown this way is eligible to be branded as Bat Friendly™. Bat conservationists, restoration biologists, and social scientists are also working with communities in the United States and Mexico to protect existing agave habitat, restore and plant new agaves, and provide educational and economic support for communities that rely on agave. You can learn more about the Tequila Interchange Project and Bat Friendly™ tequila at BatFriendly.org.

# Bats and Pesticides

With a rise in agricultural production has come an increase in the use of pesticides and chemical methods for controlling insects, fungi, bacteria, and unwanted weeds. Bats may be particularly susceptible to pesticide exposure, especially the insect-eating species that frequently forage in croplands.

The organochlorine pesticide DDT is most known for its role in the decline of raptor populations; this persistent compound moves up the food chain and has the greatest impact on top predators. However, DDT was also linked to bat population declines in the mid-1900s, such as the well-known bat colonies that inhabited Carlsbad Caverns National Park in New Mexico. Now banned in the United States, residual DDT continues to be detected in insect-eating bats around the world. DDT is thought to have contributed to more recent declines in the southern bent-wing bat in Australia and was identified as the source of a mass mortality event of big brown bats in Montana as recently as 2018.

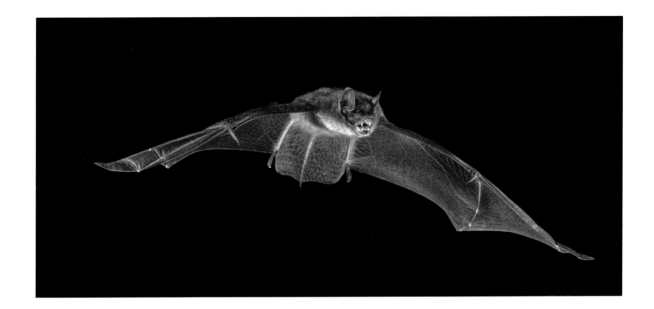

Elevated levels of pollutants such as DDT and polychlorinated biphenyls hurt bat populations already in decline due to white-nose syndrome. These compounds reduce both their immunity and the fat stores bats need to make it through the winter. Pesticides can also make it more difficult for bats to hunt. In one study, big brown bats exposed to environmentally realistic doses of the organophosphate insecticide chlorpyrifos made more mistakes when navigating a Y-maze and showed reduced activity in proteins involved in memory, learning, and sound perception. The neonicotinoid insecticide imidacloprid has also been shown to interfere with bat spatial memory by causing cell death in areas of the brain associated with memory and echolocation.

▲ Insect-eating bats like this big brown bat experience negative effects from pesticides, both in terms of reduced food resources and poisoning.

# Bat Hunting

Worldwide, bats are hunted for a variety of reasons, including use as medicine, for sport, and due to a fear that they spread disease. However, the most widespread reason for bat hunting is for consumption, both as cultural and traditional fare and as an alternative source of protein where affordability or accessibility of other types of meats is limited. At least 19 percent of bat species are estimated to experience conservation threats from hunting, a disproportionate number of which are flying foxes. Specific hunting practices vary by region, with some areas experiencing some level of hunting and/or poaching year-round, whereas other areas have specific (if not well-enforced) hunting seasons. Bat hunting practices and customs are deeply ingrained in some cultures, with archaeological evidence of bats being hunted by people as a source of protein on some islands. Economic, societal, and natural disasters that impact access to food sources also play roles in patterns of bat hunting, consumption, and trade.

From a conservation perspective, solutions to bat hunting must be regionally specific and culturally sensitive. In some regions, bat hunting persists even with regulatory laws, as the laws are either confusing, have loopholes that are easy to exploit, or are challenging to enforce in remote areas due to lack of support. In one study on bat hunting in Malaysian Borneo, residents recognized the ecological benefits of bats but did not feel that bat hunting was a conservation threat, mistakenly thinking bats can reproduce quickly. In outreach programs where hunters have joined biologists on bat population counts, the hunters frequently overestimate population sizes and are genuinely surprised when counted populations are only in the hundreds or thousands. Education and awareness programs that engage the local community in the process of bat conservation through citizen science and capacity building with local rangers have been proven to help develop sustainable management practices for bats in the Philippines (Filipinos for Flying Foxes), the Comoros Islands, and sub-Saharan Africa (Eidolon Monitoring Network).

Harvesting bats for tourist souvenirs or as macabre, gothic décor is also a rising concern, particularly in online spaces. While a simple search for "bat décor" on websites like Etsy mostly brings up the typical wares like art, jewelry, or textiles, it worryingly also brings up results like "real framed bat in shadow box" or "actual bat specimen hanging in glass dome." Often, these bat specimens being sold as decorations come with disclaimers such as "ethically sourced," "farmed," or "not endangered species," which may be outright lies. Bats are small, long-lived creatures with relatively few predators, so it's rare to encounter naturally dead bats in the wild in numbers that could support an online business. With a few exceptions, bats also don't reproduce quickly or very well in captivity, so "farming" bats for harvest is not realistic. Finally, many of these specimens come from bat biodiversity hotspots such as Southeast Asia and tropical Africa,

where population trends or conservation status are often unknown. As the rise of e-commerce platforms like Etsy, eBay, and Amazon is relatively new, the exact impact that this trade might have on bat populations is still unknown.

▲ Mauritian flying foxes are targeted as the cause of damage to fruit crops in the country, despite evidence to the contrary, leading to politically motivated culls over the past several years.

## Just a Bad Attitude?

Some people just don't like bats. Maybe they had a bad experience with a bat in their past or heard stories from loved ones who had. Maybe the leathery wings and scuttling movement of bats reminds them too much of some horror movie scene. Or maybe they lost a portion of their monthly income due to some rogue bats. With increasing encroachment on wild habitats from agriculture, livestock, and urban growth comes increased potential for bat-human conflicts.

While bats provide a great number of economically and ecologically valuable services, they also provide disservices as well.

Fruit bats, particularly flying foxes in areas of Africa, Asia, and Australia, can cause damage to commercial and small-scale fruit crops. This threat to people's livelihoods by bats—whether real or perceived—can understandably lead to negative attitudes toward bats. Unfortunately, methods of dealing with these conflicts include culling by trapping or poisoning of whole caves or roosts of bats, regardless of species. The culls themselves are rarely effective at addressing farmers' concerns, whether its rabies or crop raiding. These situations highlight the importance of good communication and capacity building between scientists, conservation managers, and local stakeholders.

Outside of direct conflict between bats and people, there are a variety of social and cultural reasons that can cause people to be disgusted by or fear bats. Culturally, bats are often associated with spirits of the dead, and they have been associated with devils and witchcraft in Christian writings dating to the four-teenth century. On the other hand, many Asian and Pacific cultures generally associate bats with positive values such as luck and blessings or as a symbol for bravery and coming of age. In one Samoan legend, flying foxes were the heroes of the story when the bats rescued Leutogi, the Tongan king's Samoan wife.

Again, education and positive exposure to bats can help moderate people's negative attitudes toward them. In one study, showing people cute pictures of bats or cartoons about how bats are helpful led to less negative feelings about bats than recorded in subjects not shown those things. Likewise, virtual reality technology was used at the Jersey Zoo in the United Kingdom to immerse visitors in a short presentation with videos and images of zoo-housed bats to increase the positive feelings people had toward bats.

# Climate Change

Bats' abilities to move and disperse via flight may make some species more resilient to changing climate conditions by facilitating range shifts or movements to different habitats. Warming winters may be facilitating the northward expansion of the adaptable Mexican free-tailed bat, whose range is constricted by cold winter temperatures. Since 2007, researchers have noted a northward expansion of the species' usual distribution in Florida, Texas, and the Southwest, with year-round colonies being established in North Carolina, Tennessee, and even Virginia. However, climate change can be a double-edged sword. Thousands of overwintering free-tailed bats roosting in bridges and urban structures perished when winter storm Uri descended on Texas in February 2021. Mortality from the storm is still unclear, though at least 10,000 bats were estimated to have died in the Houston area alone.

Most evidence suggests that climate change is going to be bad for bats. Species that are already vulnerable, such those living on islands or whose population numbers are already very low, are expected to be disproportionately affected. The increasing frequency and severity of hurricanes and cyclones could lead to devasting losses of bats, especially as some species have already experienced declines from past storms. Extreme heat waves and drought can also put species under extreme stress. Bats don't have sweat glands, and unrelenting high temperatures (above 42°C, 108°F) can cause heat stress and death from overheating, especially in larger bats like flying foxes. Climate change is also linked to more frequent and intense wildfires, which can contribute to habitat loss for forest-dwelling bats. The 2019–2020 megafires that burned more than 100,000 ha of forest in eastern Australia also led to significant loss of winter habitat for the grey-headed flying fox.

Climate change also leads to mismatches between seasonal movements of bats and the availability of cyclical resources like flowering or fruiting plants or insect emergences. A 2019 study reported that the timing overlap between the migratory Mexican long-nosed bat and agave blooms could be reduced by as much as 75 percent, depending on the future climate scenario. This would greatly reduce the bat's foraging habitat, leading to reduced *Agave* populations from lack of pollination, leading to further loss of the bat from some areas.

Addressing a global problem like climate change is going to take global solutions, like real commitments to reduce greenhouse gas emissions. That said, local and regional efforts to support bats and other species may be able to help ameliorate the effects of climate change in some cases. As global warming will contribute to ongoing habitat loss, protecting and restoring forests, wetlands, and prairies can help provide critical foraging and roosting habitat for bats. In areas prone to extreme heat events or drought, providing additional water sources as well as protecting existing rivers and wetlands could also help.

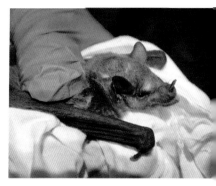

▲ This endangered Mexican long-nosed bat was captured at a maternity colony in Big Bend National Park. Pregnant females rely on blooming agave plants during their northward migration.

► Migrating species like the hoary bat are disproportionately killed by wind turbines across North America.

# Renewable Energy Development

As of 2013, more than 500,000 bats were estimated to be killed annually from wind farms in Canada and the United States. With increasing wind development, it's likely this number has only increased in the past decade. Bats either die by colliding with the turbines—which at the tip are moving at more than 200 miles per hour—or from trauma associated with rapid pressure changes near the turbines. Both migratory bat species and those adapted to flying and foraging high in open air are generally most affected by wind farms in North America and Europe. Some estimates suggest that fatalities from wind energy facilities could cause hoary bat populations to show a 50 percent decline by 2028 or to decline by as much as 90 percent in the next 50 years. Although wind energy development is increasing globally, few studies have evaluated the effects on bats in other parts of the world. As of 2021, only four published studies specifically investigated bat mortality at wind facilities in Latin America and only one published study was conducted in Asia. Clearly, scientists need to continue to evaluate the risks of wind energy in these bat-diverse regions.

Bats fatalities can be reduced by as much as 90 percent via operation mitigation and curtailment. Each type of turbine has a recommended cut-in speed, which is the wind speed at which the turbine begins generating electricity. Increasing the cut-in speed by just 1 or 2 m per second and keeping turbines rotating slowly below this threshold has been demonstrated to significantly decrease bat mortality. Unfortunately, this also reduces the amount of power and profit generated by a turbine, making it challenging to convince companies to widely adopt this strategy. Other potential methods for reducing bat mortality include the use of acoustic deterrents, devices that emit ultrasonic wavelengths that annoy or disrupt bat echolocation. So far, however, the effectiveness of these deterrents has been mixed. Some research has shown reduced mortality for some bat species but not others, whereas other studies indicated combining curtailment and deterrents could reduce bat mortality by more than just curtailment alone. Other disadvantages of acoustic deterrents are that they can be expensive to install and maintain and the deterring signal may not cover a very large area relative to the turbine.

# City Bats

Globally, the overall impact of urbanization on bats is mostly negative, though the impacts depend on bat species and the extent of development. Bats that live in open natural habitats are more likely to tolerate or benefit from urban life, as are species willing to forage around streetlights. Buildings, bridges, and other human-made structures can also mimic the natural roosts of some species, such as some cavity-roosting bats. Meanwhile, species that are obligate cave or tree roosters tend to respond poorly to urban expansion. As with much of what we know about bats, there are still large knowledge gaps of bat responses to urbanization in regions like Africa, Southeast Asia, and Central and South America.

Light and noise from even modest levels of development can also have potentially negative effects on bats. Noise can both mask important sound cues and act as distractions. For example, noise from traffic and gas compressors has been shown to significantly reduce hunting success in the pallid bat, a species that relies on listening for prey-generated sounds on the ground. When exposed to noise, pallid bats also took two to three times longer to locate the sounds of prey and increased their own calls by as much as 25 percent.

Light pollution, or artificial lighting at night, is any change in natural light levels in nocturnal systems. Some species, such as horseshoe bats, are highly sensitive to artificial lighting. Lesser horseshoe bats significantly reduced movement in areas with street lighting, while greater horseshoe bats completely changed their activity patterns to avoid lit areas entirely. By disrupting commuting areas, lighting can force bats to travel farther to reach foraging sites or reduce connectivity between social groups or populations. And, although some bats may benefit from high densities of insects drawn to streetlights, artificial lighting at night could also change the abundance and distribution of insect prey at the landscape scale. Light pollution also impacts bat behaviors at roosts, delaying emergence or resulting in the avoidance or abandonment of roosts altogether. Most research on artificial noise and lighting has focused on temperate and insect-eating bats, and it remains unclear how fruit- and nectar-feeding bats are affected.

# Making Space for Bats: Bat Boxes

For bats that tolerate urbanization and roost in human-made structures, there can still be downsides to city living—namely, an increase in human-bat conflict when bats roost in areas actively used or frequented by humans, such as houses, churches, or stadiums. Careless or purposefully cruel exclusions of bats from human spaces can result in large numbers of bats being trapped or dying. Even with careful and humane exclusions, evicting bats in an increasingly urban space with little access to natural roosts can leave them homeless. One commonly proposed way to mitigate these potential impacts or to support bats in the face of decreasing natural roosts is by hanging bat boxes. But do bat boxes really work?

As a compensatory strategy for when bats are excluded from an existing roost, the value of bat boxes is mixed. In the United Kingdom, all bats are protected under law and any construction or modification to buildings where bats are roosting requires some kind of mitigation plan. Here, the most successful mitigation technique was the creation of bat lofts, dedicated spaces sealed from the rest of the house or structure that still allow bats access from the outside. In one UK study, when the only mitigation method was bat box installation, only about 31 percent of sites retained bats.

Outside of their use as compensation for the loss of specific roosts, bat boxes have a surprisingly long history in conservation circles. In the early 1800s, some European conservationists had proposed the use of bat boxes in forests to increase the numbers of forest-pest eating bats, though the idea was generally unpopular. In the early 1900s, Dr. Charles Campbell, a physician in San Antonio, Texas, had the idea of building boxes to attract bats as a way of solving the

malaria problem caused by mosquitoes. This project was ultimately unsuccessful, as bats never occupied the boxes, but one of these structures, the Hygieostatic Bat Roost, is still standing.

Bat boxes can and do provide important roosting habitat for a broad array of bats, though only a subset of species regularly use bat boxes to raise young. Species that are already urban tolerant are more likely to use bat boxes as roosts than more urban-sensitive or specialized bat species. Some studies demonstrated positive effects of bat boxes for certain species, especially in areas where natural habitat has been lost. The Florida bonneted bat, an endangered species found only in a few areas of the state, has readily occupied specially made bat boxes, both increasing roosting resources for this bat and helping raise awareness of bat conservation among the public. Well-placed bat boxes can also attract bats to suburban or mixed agricultural areas like farms, where bats get the benefit of increased roosting habitat and humans benefit from free ecosystem services like pest control. Continued research and monitoring of the use of bat boxes by various species is needed to maximize the conservation impacts of bat boxes, especially in less well studied regions of the world.

# Maximizing your bat box success

Putting up a bat box can be a great way to help support bats in your local area, especially when paired with other bat-friendly habits. Here are some tips and tricks that can help make your bat house a success.

## Research your local bats.

Not all bat species readily roost in bat boxes, and different species have different preferences. Consider aspects of your target species' biology, like average colony size and where they naturally roost.

## Decide on a bat box style.

Once you have learned a bit about your local bats, choose a bat house style that will best match their preferences. A common, versatile style is the three-chambered bat box, which is usually rectangular and has slats inside that separate the box into three areas. Multichambered boxes are more likely to be successful that a single-chambered box, and so-called rocket boxes can be good for bats that prefer more crevice-style roosts. If you want to build your own, Bat Conservation International has construction plans for both styles available for free on their website (BatCon.org/about-bats/bat-gardens-houses/). If you'd rather buy a premade bat box, you can find bat-biologist-approved vendors at MerlinTuttle.org/selecting-a-quality-bat-house/.

## Paint it up.

Painted bat boxes are more likely to be occupied than unpainted boxes. Darker colors are recommended the closer you are to the poles (such as northern areas of North America), whereas those in more southern or hot climates should be painted lighter colors. Making sure your bat house is well ventilated and receives the right amount of sun for your local bats will also ensure that they don't overheat in their new home.

## Location, location, location.

In situations where bats are being excluded from a building roost, several studies found that mounting the bat box on the side of the building that was the original roost can increase the chance of bat occupancy. Bat houses can also be mounted on stakes or poles. Make sure the bat box is a decent height off the ground (the recommended minimum is at least 13 feet), with no obstacles below the box. Don't mount your box on a tree or close to a forest edge, as studies show these tend to be avoided.

## The more the merrier.

If you can, mounting a few bat boxes back to back or near each other will increase the chance of bat occupation. Having a few options lets bats move between the boxes based on their physiological needs or as environmental conditions change, promoting natural social interactions.

## Have patience.

Even if everything about your new bat house is perfect, it might take a while for bats to move in. In most cases, it takes bats at least a year to find a new bat box, if not longer. The local prevalence of natural roosts can influence how likely bats are to occupy a new roost. While you wait for bats to find your new box, it's also recommended that you make periodic checks and clear out any potentially unwelcome competitors such as wasps or ants.

# Conclusion

It's easy to consider bats in the abstract. Small, flying animals that we almost never see in our daily lives. If anything, I hope this book has reminded you to look past the drab, black cartoonish bat and peer behind the curtain to the true magic of bats. From bats that make their own little castles out of termite mounds to those that hunker down in the snow, every year brings new and exciting insight into the lives of bats. And just when we think we understand why or how bats do something, they frequently turn around and surprise us.

Science begins with observation. Reports of unusual behaviors like bats hibernating in the snow could have gone unnoticed, assumed to be a simple anomaly or just a misunderstood rumor. But instead of ignoring the reports, scientists spent hours trying to track down bats in the snow and learned that this behavior was more common than it seemed, revealing a whole new aspect of bat biology for continued exploration.

▶ A Rafinesque's big-eared bat comes down for a drink of water.

After the observation comes the questions. When I first started researching bats as an undergraduate, I noticed that for each major discovery scientists made about how and why bats do something, many more questions emerged that had yet to be answered. New technologies continue to help pull back the curtain a little more each year. This includes advances in gene sequencing that make it easier and faster to get down to the DNA level of a bat and the development of smaller and lighter GPS tags that let us follow bats remotely and learn where they go.

Let yourself be surprised by bats. Even after rigorous study, scientists frequently prove themselves wrong, learning new things in the process. Take for example the assumptions that leaf-nosed bats only echolocate out of their noses. If it weren't for someone sitting down and testing that idea, we would still think this was true, even though the evidence now suggests its more complicated than that. As Brock Fenton, emeritus professor of biology at Western University in Ontario, likes to say, "The bats just didn't read the research proposal!"

The facts in this book are just the beginning. There are still many more things to learn and discover about bats.

## Everyone Can Help Bats

Bats are important pollinators, seed dispersers, pest hunters, and ecological keystone species, proving that humans need bats as much as bats need us. Between climate change, urban and agricultural development, and human population expansion, the planet is quickly changing, and bats around the world face a multitude of challenges. Solving the many different conservation threats bat face is going to take creative and evidence-based solutions, engaging with communities, and working together. Even without the job title of scientist, there are many ways that you can help conserve bats and contribute to our knowledge of bats around the world.

### Create healthy bat habitat.

Bat boxes are just one way to provide new roosting habitat for bats, especially in areas where roosts are scarce or bats are being excluded from an existing roost.

Other ways to create healthy habitat include growing native plants in your yards and gardens, reducing or stopping pesticide use, and providing water features to support thirsty bats.

▲ A Jamaican fruit-eating bat roosts in an abandoned building in Panama.

## Volunteer for bats.

To make informed policies around bat conservation and management, scientists need good data on bat populations. With scientists and local communities working together, we can learn so much more than we can by ourselves. Check in with your local conservation or land preservation organizations or state and county parks to see if there are opportunities to count bats at roosts or assist with bat acoustic monitoring.

## Advocate for bats.

While public opinion in many places is shifting when it comes to bats, there are still people who fear or hate bats. When you hear people spreading myths about bats or see media reports that promote potentially cruel treatment of bats, speak up! Even if people don't love bats the way you or I do, it's important to promote the respect and care for bats as living creatures and important parts of the world's ecosystems.

# Cast of Bats

(Remember that bats do not have standardized common names, so you may sometimes see variation in common names assigned to different species.)

| Family | Common name | Scientific name |
|---|---|---|
| **Craseonycteridae (hog-nosed bats)** | Kitti's hog-nosed bat | *Craseonycteris thonglongyai* |
| **Emballonuridae (sac-winged bats)** | northern ghost bat | *Diclidurus albus* |
| | proboscis bat | *Rhynchonycteris naso* |
| | greater sac-winged bat | *Saccopteryx bilineata* |
| | lesser sac-winged bat | *Saccopteryx leptura* |
| | long-winged tomb bat | *Taphozous longimanus* |
| **Hipposideridae (Old World leaf-nosed bats)** | Percival's trident bat | *Cloeotis percivali* |
| | diadem leaf-nosed bat | *Hipposideros diadema* |
| | Schneider's leaf-nosed bat | *Hipposideros speoris* |
| **Megadermatidae (false vampire bats)** | heart-nosed bat | *Cardioderma cor* |
| | yellow-winged bat | *Lavia frons* |
| | Australian ghost bat | *Macroderma gigas* |
| | greater false-vampire bat | *Megaderma lyra* |
| **Miniopteridae (bent-winged bats)** | southern bent-winged bat | *Miniopterus orianae* |
| | common bent-winged bat | *Miniopterus schreibersii* |
| **Molossidae (free-tailed and mastiff bats)** | Fijian free-tailed bat | *Chaerephon bregullae* |
| | Chapin's free-tailed bat | *Chaerephon chapini* |
| | wrinkle-lipped free-tailed bat | *Chaerephon plicatus* |
| | Florida bonneted bat | *Eumops floridanus* |
| | velvety free-tailed bat | *Molossus molossus* |
| | black mastiff bat | *Molossus nigricans* |
| | Sinaloan mastiff bat | *Molossus sinaloae* |
| | Mexican free-tailed bat | *Tadarida brasliensis* |
| | European free-tailed bat | *Tadarida teniotis* |

| Family | Common name | Scientific name |
| --- | --- | --- |
| **Mormoopida (mustached bats)** | ghost-faced bat | *Mormoops megalophylla* |
| | tawny naked-backed bat | *Pteronotus fulvus* |
| | Mesoamerican mustached bat | *Pteronotus mesoamericanus* |
| | Parnelli's mustached bat | *Pteronotus parnellii* |
| **Mystacinidae (New Zealand bats)** | New Zealand lesser short-tailed bat | *Mystacina tuberculata* |
| **Myzopodidae (sucker-footed bats)** | Madagascar sucker-footed bat | *Myzopoda aurita* |
| **Natalidae (funnel-eared bats)** | Mexican funnel-eared bat | *Natalus mexicanus* |
| **Noctilionidae (bulldog bats)** | lesser bulldog bat | *Noctilio albiventris* |
| | greater bulldog bat | *Noctilio leporinus* |
| **Nycteridae (slit-faced bats)** | Egyptian slit-faced bat | *Nycteris thebaica* |
| **Phyllostomidae (New World leaf-nosed bats)** | tailed tailless bat | *Anoura caudifer* |
| | tube-lipped nectar bat | *Anoura fistulata* |
| | Geoffroy's tailless bat | *Anoura geoffroyi* |
| | Jamaican fruit-eating bat | *Artibeus jamaicensis* |
| | great fruit-eating bat | *Artibeus lituratus* |
| | Cuban fruit-eating bat | *Brachyphylla nana* |
| | Seba's short-tailed fruit bat | *Carollia perspicillata* |
| | Sowell's short-tailed fruit bat | *Carollia sowelli* |
| | wrinkle-faced bat | *Centurio senex* |
| | hairy big-eyed bat | *Chiroderma villosum* |

| Family | Common name | Scientific name |
|---|---|---|
| **Phyllostomidae (New World leaf-nosed bats [continued])** | Mexican long-tongued bat | *Choeronycteris mexicana* |
| | woolly false-vampire bat | *Chrotopterus auritus* |
| | Thomas's fruit-eating bat | *Dermanura watsoni* |
| | common vampire bat | *Desmodus rotundus* |
| | white-winged vampire bat | *Diaemus youngi* |
| | hairy-legged vampire bat | *Diphylla ecaudata* |
| | Honduran white tent-making bat | *Ectophylla alba* |
| | striped hairy-nosed bat | *Gardnernycteris crenulatum* |
| | Merriam's long-tongued bat | *Glossophaga mutica* |
| | Pallas's long-tongued bat | *Glossophaga soricina* |
| | southern long-nosed bat | *Leptonycteris curasoae* |
| | Mexican long-nosed bat | *Leptonycteris nivalis* |
| | lesser long-nosed bat | *Leptonycteris yerbabuenae* |
| | orange nectar bat | *Lonchophylla robusta* |
| | Davis's round-eared bat | *Lophostoma evotis* |
| | white-throated round-eared bat | *Lophostoma silvicolum* |
| | common big-eared bat | *Micronycteris microtis* |
| | Cozumelan golden bat | *Mimon cozumelae* |
| | Jamaican flower bat | *Phyllonycteris aphylla* |
| | pale spear-nosed bat | *Phyllostomus discolor* |
| | greater spear-nosed bat | *Phyllostomus hastatus* |
| | white-lined broad-nosed bat | *Platyrrhinus lineatus* |
| | little yellow-shouldered bat | *Sturnira lilium* |
| | northern yellow-shouldered bat | *Sturnira parvidens* |
| | Tilda's yellow-shouldered bat | *Sturnira tildae* |
| | Peter's tent-making bat | *Uroderma bilobatum* |
| | Pacific tent-making bat | *Uroderma convexum* |
| | spectral bat | *Vampyrum spectrum* |

| Family | Common name | Scientific name |
| --- | --- | --- |
| **Pteropodidae (Old World fruit bats and flying foxes)** | lesser short-nosed fruit bat | *Cynopterus brachyotis* |
| | greater short-nosed fruit bat | *Cynopterus sphinx* |
| | Dayak fruit bat | *Dyacopterus spadiceus* |
| | straw-colored fruit bat | *Eidolon helvum* |
| | cave nectar bat | *Eonycteris spelaea* |
| | Wahlberg's epauletted fruit bat | *Epomophorus wahlbergi* |
| | Fischer's pygmy fruit bat | *Haplonycteris fischeri* |
| | hammer-headed bat | *Hypsignathus monstrosus* |
| | northern blossom bat | *Macroglossus minimus* |
| | black flying fox | *Pteropus alecto* |
| | spectacled flying fox | *Pteropus conspicillatus* |
| | Ryukyu flying fox | *Pteropus dasymallus* |
| | Indian flying fox | *Pteropus giganteus* |
| | island flying fox | *Pteropus hypomelanus* |
| | Christmas Island flying fox | *Pteropus natalis* |
| | Bismarck flying fox | *Pteropus neohibernicus* |
| | Mauritian flying fox | *Pteropus niger* |
| | grey-headed flying fox | *Pteropus poliocephalus* |
| | Rodrigues fruit bat | *Pteropus rodricensis* |
| | little red flying fox | *Pteropus scapulatus* |
| | Egyptian fruit bat | *Rousettus aegyptiacus* |
| | common blossom bat | *Syconycteris australis* |
| **Rhinolophidae (horseshoe bats)** | eloquent horseshoe bat | *Rhinolophus eloquens* |
| | greater horseshoe bat | *Rhinolophus ferrumequinum* |
| | lesser horseshoe bat | *Rhinolophus hipposideros* |
| | least horseshoe bat | *Rhinolophus pusillus* |
| | little Japanese horseshoe bat | *Rhinolophus cornutus* |

| Family | Common name | Scientific name |
|---|---|---|
| **Rhinolophidae (horseshoe bats [continued])** | rufous horseshoe bat | *Rhinolophus rouxii* |
| | trefoil horseshoe bat | *Rhinolophus trifoliatus* |
| **Rhinopomatidae (mouse-tailed bats)** | greater mouse-tailed bat | *Rhinopoma microphyllum* |
| **Thyropteridae (disc-winged bats)** | Spix's disc-winged bat | *Thyroptera tricolor* |
| **Vesptertilionidae (evening bats)** | banana serotine | *Afronycteris nana* |
| | pallid bat | *Antrozous pallidus* |
| | barbastelle bat | *Barbastella barbastelle* |
| | Van Gelder's bat | *Bauerus dubiaquercus* |
| | Gould's wattled bat | *Chalinolobus gouldii* |
| | Rafinesque's big-eared bat | *Corynorhinus rafinesquii* |
| | Townsend's big-eared bat | *Corynorhinus townsendii* |
| | Virginia big-eared bat | *Corynorhinus townsendii virginianus* |
| | big brown bat | *Eptesicus fuscus* |
| | northern bat | *Eptesicus nilsonii* |
| | spotted bat | *Euderma maculatum* |
| | Hardwicke's woolly bat | *Kerivoula hardwickii* |
| | golden-tipped bat | *Kerivoula papuensis* |
| | painted bat | *Kerivoula picta* |
| | silver-haired bat | *Lasionycteris noctivagans* |
| | western red bat | *Lasiurus blossevillii* |
| | eastern red bat | *Lasiurus borealis* |

| Family | Common name | Scientific name |
|---|---|---|
| **Vesptertilionidae (evening bats [*continued*])** | hoary bat | *Lasiurus cinereus* |
| | southern yellow bat | *Lasiurus ega* |
| | northern yellow bat | *Lasiurus intermedius* |
| | Ussuri tube-nosed bat | *Murina ussuriensis* |
| | silver-tipped myotis | *Myotis albescens* |
| | southeastern myotis | *Myotis austroriparius* |
| | Bechstein's myotis | *Myotis bechsteinii* |
| | Brandt's myotis | *Myotis brandtii* |
| | California myotis | *Myotis californicus* |
| | long-fingered myotis | *Myotis capaccinii* |
| | Daubenton's myotis | *Myotis daubentonii* |
| | David's myotis | *Myotis davidii* |
| | grey bat | *Myotis grisescens* |
| | hairy-legged myotis | *Myotis keaysi* |
| | eastern small-footed bat | *Myotis leibii* |
| | mouse-eared bat | *Myotis myotis* |
| | Natterer's bat | *Myotis nattereri* |
| | Nimba Mountain myotis | *Myotis nimbaensis* |
| | northern long-eared bat | *Myotis septentrionalis* |
| | Indiana bat | *Myotis sodalis* |
| | cave myotis | *Myotis velifer* |
| | Mexican fishing bat | *Myotis vivesi* |

| Family | Common name | Scientific name |
|---|---|---|
| **Vesptertilionidae (evening bats [continued])** | Yuma myotis | *Myotis yumaensis* |
| | greater noctule bat | *Nyctalus lasiopterus* |
| | lesser noctule | *Nyctalus leisleri* |
| | common noctule bat | *Nyctalus noctula* |
| | Chinese noctule | *Nyctalus plancyi* |
| | evening bat | *Nycticeius humeralis* |
| | desert long-eared bat | *Otonycteris hemprichii* |
| | tricolored bat | *Perimyotis subflavus* |
| | Kuhl's pipistrelle | *Pipistrellus kuhlii* |
| | common pipistrelle | *Pipistrellus pipistrellus* |
| | soprano pipistrelle | *Pipistrellus pygmaeus* |
| | brown long-eared bat | *Plecotus auritus* |
| | lesser bamboo bat | *Tylonycteris pachypus* |
| | greater bamboo bat | *Tylonycteris robustula* |
| | southern forest bat | *Vespadelus regulus* |
| | parti-colored bat | *Vespertilio murinus* |

# Acknowledgments

I'd like to start by thanking Dr. Brock Fenton, Sherry Fenton, and Price Sewell for allowing me to use their stunning bat photography throughout much of this book. Thank you also to NA Bat and Christen Long for providing access to bat echolocation call recordings. I'd like to thank Dr. Winifred Frick for thinking of me when the opportunity to write this book was presented to her. Thanks also to Julia Brokaw for reviewing many of the chapters of this book.

I'd also like to extend my gratitude to the many people who have made my scientific work with bats so rewarding: Amanda Adams, Kushal Bakshi, Dan Becker, David Boerma, Jeff Clerc, Evynn Davis, Jon Flanders, Fran Hutchins, Melissa Ingala, Silvio Macias, Brian O'Toole, Rachel Page, Kyle Shute, Nancy Simmons, Grace Smarsh, Mike Smotherman, Kelly Speer, Joe Szewczak, Samantha Trent, Ted Weller, Amy Wray, and the many others who have provided thoughtful discussions, advice, and camaraderie throughout the years.

Finally, thanks to Brad Bogdan for your endless enthusiasm, support, and love both during the writing of this book and throughout my career. I love you.

# Photo and Illustration Credits

## Photographs

*All photographs are by the author except the following:*

Brock Fenton, 15, 23 (left), 25, 27, 31, 38 (middle, right), 39, 41, 48, 51, 54 (top), 63, 64–65, 71, 73, 78, 80, 87, 93, 95, 98, 100, 101, 103, 105 (left), 110, 111, 113, 126, 127 (bottom), 128 (left), 129, 132, 133 (top, middle), 137 (bottom), 139 (right), 146, 147, 151, 157, 159, 167, 170, 187 (right), 190, 197, 201, 203–209, 213, 218 (top), 220, 232 (right), 235

Grant Maslowski/Nature Picture Library, 112

Jon Flanders/Bat Conservation International, 13, 105 (right)

Melqui Gamba-Rios/Bat Conservation International, 142, 223

©MerlinTuttle.org, 164

Price Sewell, 18, 43, 47, 52, 57, 81, 90–91, 94, 97, 107, 114, 131, 133 (bottom), 141, 153 (left), 163, 168, 187 (left, middle), 188, 193, 220 (left), 226, 227 (right), 228, 238 (top, bottom), 247

Rachel Harper/Bat Conservation International, 22 (right)

Rietbergen TB, van den Hoek Ostende LW, Aase A, Jones MF, Medeiros ED, Simmons NB (2023) The oldest known bat skeletons and their implications for Eocene chiropteran diversification. PLoS ONE 18(4): e0283505. 99

Scan-bugs.org/Yale Peabody Museum, 82

Taylor PJ, Stoffberg S, Monadjem A, Schoeman MC, Bayliss J, Cotterill FPD (2012) Four New Bat Species (Rhinolophus hildebrandtii Complex) Reflect Plio-Pleistocene Divergence of Dwarfs and Giants across an Afromontane Archipelago. PLoS ONE 7(9): e41744. 185

Winifred Frick/Bat Conservation International, 218 (bottom)

## Flickr
**Public Domain**
Ann Froschauer/USFWS, 120–121
**CC-BY-SA 2.0**
Ann Froschauer/USFWS, 149 (right)
Christina Butler, 128 (right)
Peter Paplanus, 214

## iNaturalist
**CC-BY-SA 4.0 International**
Rolf Lawrenz, 138

## iStock
Bedebi, 172
Goddard_Photography, 169
MalcolmB2, 36

## Wikimedia
**Public Domain**
Portrait of Lazzaro Spallanzani, before 1842. Engraving by Caterina Piotti-Pirola. Archivio storico dell'Accademia delle Scienze di Torino, 26
**CC BY 2.0**
Gufm, 233
**CC BY 3.0**
CSIRO, 60 (top, right)
**CC-BY-SA 2.0**
Telegro, 20

**CC-BY-SA 3.0**
Cheers!, 219
Forest and Kim Star, 33
Marco Schmidt, 54 (middle)
**CC-BY-SA 4.0 International**
Andrew Mercer, 74, 218 (middle)
Charles J Sharp, 237
Dave Hemprich-Bennett, 32
John Robert McPherson, 155
Konrad Bidziński, 183
Luke Marcos Imbong, 195
Manojiritty, 221 (right)
Peter Addor, 175
Steve Bourne, 161 (right)
Uwe Schmidt, 109

## Illustrations

Melissa Ingala, 16, 21, 28, 30, 45, 59, 96

# References

## Introduction

Simmons, Nancy B., Jon Flanders, Eric Moïse Bakwo Fils, Guy Parker, Jamison D. Suter, Seinan Bamba, Mory Douno, Mamady Kobele Keita, Ariadna E. Morales, and Winifred F. Frick. 2021. A New Dichromatic Species of *Myotis* (Chiroptera: Vespertilionidae) from the Nimba Mountains, Guinea. *American Museum Novitates* 2020(3963):1–40.

## 1. Origins

Ammerman, Loren K., and David M. Hillis. 1992. A Molecular Test of Bat Relationships: Monophyly or Diphyly? *Systematic Biology* 41(2):222.

Jones, Gareth, and Emma C. Teeling. 2006. The Evolution of Echolocation in Bats. *Trends in Ecology & Evolution* 21(3):149–156.

Teeling, Emma C., Ole Madsen, Ronald A. Van den Bussche, Wilfried W. de Jong, Michael J. Stanhope, and Mark S. Springer. 2002. Microbat Paraphyly and the Convergent Evolution of a Key Innovation in Old World Rhinolophoid Microbats. *Proceedings of the National Academy of Sciences of the United States of America* 99(3):1431–1436.

Tsagkogeorga, Georgia, Joe Parker, Elia Stupka, James A. Cotton, and Stephen J. Rossiter. 2013. Phylogenomic Analyses Elucidate the Evolutionary Relationships of Bats. *Current Biology* 23(22):2262–2267.

## 2. Echo

Boonman, Arjan, Sara Bumrungsri, and Yossi Yovel. 2014. Nonecholocating Fruit Bats Produce Biosonar Clicks with Their Wings. *Current Biology* 24:2962–2967.

Brinkløv, Signe, Elisabeth K. V. Kalko, and Annemarie Surlykke. 2009. Intense Echolocation Calls from Two "Whispering" Bats, *Artibeus jamaicensis* and *Macrophyllum macrophyllum* (Phyllostomidae). *Journal of Experimental Biology* 212(1):11–20.

Denzinger, Annette, and Hans-Ulrich Schnitzler. 2013. Bat Guilds, a Concept to Classify the Highly Diverse Foraging and Echolocation Behaviors of Microchiropteran Bats. *Frontiers in Physiology* 4(164):1–15.

Dijkgraaf,, S. 1960. Spallanzani's Unpublished Experiments on the Sensory Basis of Object Perception in Bats. *Isis* 51:9–20.

Fenton, M. Brock. 2013. Evolution of Echolocation. In *Bat Evolution, Ecology, and Conservation*. Edited by Rick A. Adams and Scott C. Pedersen. New York: Springer. 47–70.

Fenton, M. Brock. 2013. Questions, Ideas and Tools: Lessons from Bat Echolocation. *Animal Behaviour* 85(5):869–879.

Fenton, M. Brock, and James H. Fullard. 1979. The Influence of Moth Hearing on Bat Echolocation Strategies. *Journal of Comparative Physiology A* 132(1):77–86.

Galambos, Robert. 1942. The Avoidance of Obstacles by Flying Bats: Spallanzani's Ideas (1794) and Later Theories. *Isis* 34(2):132–140.

Gessinger, Gloria, Rachel Page, Lena Wilfert, Annemarie Surlykke, Signe Brinkløv, and Marco Tschapka. 2021. Phylogenetic Patterns in Mouth Posture and Echolocation Emission Behavior of Phyllostomid Bats. *Frontiers in Ecology and Evolution* 9(May):270.

Griffin, Donald R. 1951. Audible and Ultrasonic Sounds of Bats. *Experientia* 7(12):448–453.

Griffin, Donald R., Frederic A. Webster, and Charles R. Michael. 1960. The Echolocation of Flying Insects by Bats. *Animal Behaviour* 8(3–4):141–154.

Grinnell, Alan D., Edwin Gould, and M. Brock Fenton. 2016. A History of the Study of Echolocation. In *Bat Bioacoustics*. Edited by M. Brock Fenton, Alan D Grinnell, Arthur N. Popper, and Richard R. Fay. New York: Springer. 1–24.

Jakobsen, Lasse, Signe Brinkløv, and Annemarie Surlykke. 2013. Intensity and Directionality of Bat Echolocation Signals. *Frontiers in Physiology* 4(April):89.

Jones, Gareth, and Emma C. Teeling. 2006. The Evolution of Echolocation in Bats. *Trends in Ecology & Evolution* 21(3):149–156.

Lawrence, B. D., and J. A. Simmons. 1982. Echolocation in Bats: The External Ear and Perception of the Vertical Positions of Targets. *Science* 218(4571):481–483.

Lee, Wu Jung, Benjamin Falk, Chen Chiu, Anand Krishnan, Jessica H. Arbour, and Cynthia F. Moss. 2017. Tongue-Driven Sonar Beam Steering by a Lingual-Echolocating Fruit Bat. *PLoS Biology* 15(12):e2003148.

Macias, Silvio, Kushal Bakshi, and Michael Smotherman. 2020. Functional Organization of the Primary Auditory Cortex of the Free-Tailed Bat *Tadarida brasiliensis*. *Journal of Comparative Physiology A* 206(3):429–440.

Metzner, Walter, and Rolf Müller. 2016. Ultrasound Production, Emission, and Reception. In *Bat Bioacoustics*. Edited by M. Brock Fenton, Alan D Grinnell, Arthur N. Popper, and Richard R. Fay. New York: Springer. 55–91.

Metzner, Walter, Shuyi Zhang, and Michael Smotherman. 2002. Doppler-Shift Compensation Behavior in Horseshoe Bats Revisited: Auditory Feedback Controls Both a Decrease and an Increase in Call Frequency. *Journal of Experimental Biology* 205(11):1607–1616.

Moss, Cynthia F., and Shiva R. Sinha. 2003. Neurobiology of Echolocation in Bats. *Current Opinion in Neurobiology* 13(6):751–758.

Obrist, Martin K. 1995. Flexible Bat Echolocation: The Influence of Individual, Habitat and Conspecifics on Sonar Signal Design. *Behavioral Ecology and Sociobiology* 36(3):207–219.

Obrist, Martin K., M. Brock Fenton, Judith L. Eger, and Peter A. Schlegel. 1993. What Ears Do for Bats: A Comparative Study of Pinna Sound Pressure Transformation in Chiroptera. *Journal of Experimental Biology* 180:119–152.

Schnitzler, Hans Ulrich, and Annette Denzinger. 2011. Auditory Fovea and Doppler Shift Compensation: Adaptations for Flutter Detection in Echolocating Bats Using CF-FM Signals. *Journal of Comparative Physiology A* 197(5):541–559.

Schnitzler, Hans-Ulrich, Cynthia F. Moss, and Annette Denzinger. 2003. From Spatial Orientation to Food Acquisition in Echolocating Bats. *Trends in Ecology & Evolution* 18(8):386–394.

Smotherman, Michael S., Thomas Croft, and Silvio Macias. 2022. Biosonar Discrimination of Fine Surface Textures by Echolocating Free-Tailed Bats. *Frontiers in Ecology and Evolution* 10(November):1079.

Sulser, R. Benjamin, Bruce D. Patterson, Daniel J. Urban, April I. Neander, and Zhe-Xi Luo. 2022. Evolution of Inner Ear Neuroanatomy of Bats and Implications for Echolocation. *Nature* 602(7897):449–454.

Surlykke, Annemarie, and Elisabeth K. V. Kalko. 2008. Echolocating Bats Cry Out Loud to Detect Their Prey. *PLoS ONE* 3(4):e2036.

Yin, Xiaoyan, and Rolf Müller. 2019. Fast-Moving Bat Ears Create Informative Doppler Shifts. *Proceedings of the National Academy of Sciences of the United States of America* 116(25):12270–12274.

# 3. Scent

Bhatnagar, Kunwar P., and Frank C. Kallen. 1974. Cribriform Plate of Ethmoid, Olfactory Bulb and Olfactory Acuity in Forty Species of Bats. *Journal of Morphology* 142(1):71–90.

Bhatnagar, Kunwar P., and Frank C. Kallen. 1974. Morphology of the Nasal Cavities and Associated Structures in *Artibeus jamaicensis* and *Myotis lucifugus*. *American Journal of Anatomy* 139(2):167–189.

Buck, Linda, and Richard Axel. 1991. A Novel Multigene Family May Encode Odorant Receptors: A Molecular Basis for Odor Recognition. *Cell* 65(1):175–187.

Carter, Gerald G., and Alyssa B. Stewart. 2015. The Floral Bat Lure Dimethy Disulphide Does Not Attract the Palaeotropical Dawn Bat. *Journal of Pollination Ecology* 17(19):129–131.

Hayden, Sara, Michaël Bekaert, Tess A. Crider, Stefano Mariani, William J. Murphy, and Emma C. Teeling. 2010. Ecological Adaptation Determines Functional Mammalian Olfactory Subgenomes. *Genome Research* 20(1):1–9.

Hayden, Sara, Michaël Bekaert, Alisha Goodbla, William J. Murphy, Liliana M. Dávalos, and Emma C. Teeling. 2014. A Cluster of Olfactory Receptor Genes Linked to Frugivory in Bats. *Molecular Biology and Evolution* 31(4):917–927.

Helversen, O. Von, L. Winkler, and H. J. Bestmann. 2000. Sulphur-Containing "Perfumes" Attract Flower-Visiting Bats. *Journal of Comparative Physiology A* 186(2):143–153.

Hutcheon, James M., John A. W. Kirsch, Theodore Garland Jr. 2002. A Comparative Analysis of Brain Size in Relation to Foraging Ecology and Phylogeny in the Chiroptera. *Brain, Behavior and Evolution* 60(3):165–180.

Jiang, Yue, and Hiroaki Matsunami. 2015. Mammalian Odorant Receptors: Functional Evolution and Variation. *Current Opinion in Neurobiology* 34(October):54–60.

Kalko, Elisabeth K. V., Edward Allen Herre, and Charles O. Handley Jr. 1996. Relation of Fig Fruit Characteristics to Fruit-Eating Bats in the New and Old World Tropics. *Journal of Biogeography* 23(4):565–576.

Korine, Carmi, and Elisabeth K. V. Kalko. 2005. Fruit Detection and Discrimination by Small Fruit-Eating Bats (Phyllostomidae): Echolocation Call Design and Olfaction. *Behavioral Ecology and Sociobiology* 59(1):12–23.

Rieger, James F., and Elizabeth M. Jakob. 1988. The Use of Olfaction in Food Location by Frugivorous Bats. *Biotropica* 20(2):161–164.

Sánchez, Francisco, Carmi Korine, Marco Steeghs, Luc-Jan Laarhoven, Simona M. Cristescu, Frans J. M. Harren, Robert Dudley, and Berry Pinshow. 2006. Ethanol and Methanol as Possible Odor Cues for Egyptian Fruit Bats (*Rousettus aegyptiacus*). *Journal of Chemical Ecology* 32(6):1289–1300.

Sánchez, F., C. Korine, B. Kotler, and B. Pinshow. 2007. Ethanol and Sugars as Complementary Resources for Egyptian Fruit Bats (*Rousettus aegyptiacus*). *Comparative Biochemistry and Physiology Part A: Molecular & Integrative Physiology* 146(4):S173.

Schmieder, Daniela A., Tigga Kingston, Rosli Hashim, and Björn M. Siemers. 2012. Sensory Constraints on Prey Detection Performance in an Ensemble of Vespertilionid Understorey Rain Forest Bats. *Functional Ecology* 26(5):1043–1053.

Tsagkogeorga, Georgia, Steven Müller, Christophe Dessimoz, and Stephen J. Rossiter. 2017. Comparative Genomics Reveals Contraction in Olfactory Receptor Genes in Bats. *Scientific Reports* 7(1):259.

Yohe, Laurel R., Leith B. Leiser-Miller, Zofia A. Kaliszewska, Paul Donat, Sharlene E. Santana, and Liliana M. Dávalos. 2021. Diversity in Olfactory Receptor Repertoires Is Associated with Dietary Specialization in a Genus of Frugivorous Bat. *G3: Genes, Genomes, Genetics* 11(10):jkab260.

# 4. View

Buchler, E. R., and S. B. Childs. 1982. Use of the Post-Sunset Glow as an Orientation Cue by Big Brown Bats (*Eptesicus fuscus*). *Journal of Mammalogy* 53(2):243–247.

Céchetto, Clément, Lasse Jakobsen, and Eric J. Warrant. 2023. Visual Detection Threshold in the Echolocating Daubenton's Bat (*Myotis daubentonii*). *Journal of Experimental Biology* 226(2):jeb244451.

Chase, Julia. 1981. Visually Guided Escape Responses of Microchiropteran Bats. *Animal Behaviour* 29(3):708–713.

Danilovich, S., and Y. Yovel. 2019. Integrating Vision and Echolocation for Navigation and Perception in Bats. *Science Advances* 5(6):eaaw6503.

Eklöf, Johan, Jurğis Šuba, Gunars Petersons, and Jens Rydell. 2014. Visual Acuity and Eye Size in Five European Bat Species in Relation to Foraging and Migration Strategies. *Environmental and Experimental Biology* 12:1–6.

Gracheva, Elena O., Julio F. Cordero-Morales, José A. González-Carcacía, Nicholas T. Ingolia, Carlo Manno, Carla I. Aranguren, Jonathan S. Weissman, and David Julius. 2011. Ganglion-Specific Splicing of TRPV1 Underlies Infrared Sensation in Vampire Bats. *Nature* 476(7358):88–91.

Holland, Richard A., Kasper Thorup, Maarten J. Vonhof, William W. Cochran, and Martin Wikelski. 2006. Bat Orientation Using Earth's Magnetic Field. *Nature* 444(7120):702.

Holland, Richard A., Joseph L. Kirschvink, Thomas G. Doak, and Martin Wikelski. 2008. Bats Use Magnetite to Detect the Earth's Magnetic Field. *PLoS ONE* 3(2):e1676.

Holland, Richard A., Ivailo Borissov, and Björn M. Siemers. 2010. A Nocturnal Mammal, the Greater Mouse-Eared Bat, Calibrates a Magnetic Compass by the Sun. *Proceedings of the National Academy of Sciences of the United States of America* 107(15):6941–6945.

Kürten, Ludwig, and Uwe Schmidt. 1982. Thermoperception in the Common Vampire Bat (*Desmodus rotundus*). *Journal of Comparative Physiology A* 146(2):223–228.

Lindecke, Oliver, Alise Elksne, Richard A. Holland, Gunārs Pētersons, and Christian C. Voigt. 2019. Experienced Migratory Bats Integrate the Sun's Position at Dusk for Navigation at Night. *Current Biology* 29(8):1369–1373.e3.

Marcos Gorresen, P., Paul M. Cryan, David C. Dalton, Sandy Wolf, and Frank J. Bonaccorso. 2015. Ultraviolet Vision May Be Widespread in Bats. *Acta Chiropterologica* 17(1):193–198.

Mistry, Shahroukh. 1990. Characteristics of the Visually Guided Escape Response of the Mexican Free-Tailed Bat, *Tadarida brasiliensis mexicana*. *Animal Behaviour* 39(2):314–320.

Orbach, Dara N., and M. Brock Fenton. 2010. Vision Impairs the Abilities of Bats to Avoid Colliding with Stationary Obstacles. *PLoS ONE* 5(11):e13912.

Rydell, Jens, and Johan Eklöf. 2003. Vision Complements Echolocation in an Aerial-Hawking Bat. *Naturwissenschaften* 90(10):481–483.

Simões, Bruno F., Nicole M. Foley, Graham M. Hughes, Huabin Zhao, Shuyi Zhang, Stephen J. Rossiter, and Emma C. Teeling. 2019. As Blind as a Bat? Opsin Phylogenetics Illuminates the Evolution of Color Vision in Bats. *Molecular Biology and Evolution* 36(1):54–68.

Tian, Lanxiang, Wei Lin, Shuyi Zhang, and Yongxin Pan. 2010. Bat Head Contains Soft Magnetic Particles: Evidence from Magnetism. *Bioelectromagnetics* 31(7):499–503.

Veilleux, Carrie C., and E. Christopher Kirk. 2014. Visual Acuity in Mammals: Effects of Eye Size and Ecology. *Brain Behavior and Evolution* 83(1):43–53.

Wang, Daryi, Todd Oakley, Jeffrey Mower, Lawrence C. Shimmin, Sokchea Yim, Rodney L. Honeycutt, Hsienshao Tsao, and Wen-Hsiung Li. 2004. Molecular Evolution of Bat Color Vision Genes. *Molecular Biology and Evolution* 21(2):295–302.

Wang, Yinan, Yongxin Pan, Stuart Parsons, Michael Walker, and Shuyi Zhang. 2007. Bats Respond to Polarity of a Magnetic Field. *Proceedings of the Royal Society B: Biological Sciences* 274(1627):2901–2905.

Winter, York, Jorge López, and Otto Von Helversen. 2003. Ultraviolet Vision in a Bat. *Nature* 425(6958):612–614.

Yovel, Yossi, and Nachum Ulanovsky. 2017. Bat Navigation. In *Learning and Memory: A Comprehensive Reference*. Edited by John H. Byrne. Philadelphia, PA: Elsevier. 333–345.

# 5. Bite

Aizpurua, Ostaizka, and Antton Alberdi. 2018. Ecology and Evolutionary Biology of Fishing Bats. *Mammal Review* 48(4):284–297.

Barber, Jesse R., Brad A. Chadwell, Nick Garrett, Barbara Schmidt-French, and William E. Conner. 2009. Naïve Bats Discriminate Arctiid Moth Warning Sounds But Generalize Their Aposematic Meaning. *Journal of Experimental Biology* 212(14):2141–2148.

Barber, Jesse R., Brian C. Leavell, Adam L. Keener, Jesse W. Breinholt, Brad A. Chadwell, Christopher J. W. McClure, Geena M. Hill, and Akito Y. Kawahara. 2015. Moth Tails Divert Bat Attack: Evolution of Acoustic Deflection. *Proceedings of the National Academy of Sciences of the United States of America* 112(9):2812–2816.

Blumer, Moritz, Tom Brown, Mariella Bontempo Freitas, Ana Luiza Destro, Juraci A. Oliveira, Ariadna E. Morales, Tilman Schell, et al. 2022. Gene Losses in the Common Vampire Bat Illuminate Molecular Adaptations to Blood Feeding. *Science Advances* 8(12):6494.

Bobrowiec, Paulo Estefano D., Maristerra R. Lemes, and Rogério Gribel. 2015. Prey Preference of the Common Vampire Bat (*Desmodus rotundus*, Chiroptera) Using Molecular Analysis. *Journal of Mammalogy* 96(1):54–63.

Clare, Elizabeth L., Holger R. Goerlitz, Violaine A. Drapeau, Marc W. Holderied, Amanda M. Adams, Juliet Nagel, Elizabeth R. Dumont, Paul D. N. Hebert, and M. Brock Fenton. 2013. Trophic Niche Flexibility in *Glossophaga soricina*: How a Nectar Seeker Sneaks an Insect Snack. *Functional Ecology* 28(3):632–641.

Conner, William E., and Aaron J. Corcoran. 2011. Sound Strategies: The 65-Million-Year-Old Battle Between Bats and Insects. *Annual Review of Entomology* 57(1):21–39.

Corcoran, Aaron J., Jesse R. Barber, and William E. Conner. 2009. Tiger Moth Jams Bat Sonar. *Science* 325(5938):325–327.

Corcoran, Aaron J., Jesse R. Barber, Nickolay I. Hristov, and William E. Conner. 2011. How Do Tiger Moths Jam Bat Sonar? *Journal of Experimental Biology* 214(14):2416–2425.

Dechmann, Dina K. N., Kamran Safi, and Maarten J. Vonhof. 2006. Matching Morphology and Diet in the Disc-Winged Bat *Thyroptera tricolor* (Chiroptera). *Journal of Mammalogy* 87(5):1013–1019.

Dixon, M. May, Patricia L. Jones, Michael J. Ryan, Gerald G. Carter, and Rachel A. Page. 2022. Long-Term Memory in Frog-Eating Bats. *Current Biology* 32(12):R557–558.

Dondini, Gianna. 2005. Bats: Bird-Eaters or Feather-Eaters? A Contribution to Debate on Great Noctule Carnivory. *Hystrix* 15(2):86–88.

Dumont, Elizabeth R. 1999. The Effect of Food Hardness on Feeding Behaviour in Frugivorous Bats (Phyllostomidae): An Experimental Study. *Journal of Zoology* 248(2):219–229.

Dumont, E. R., A. Herrel, R. A. Medellín, J. A. Vargas-Contreras, and S. E. Santana. 2009. Built to Bite: Cranial Design and Function in the Wrinkle-Faced Bat. *Journal of Zoology* 279(4):329–337.

Duque-Márquez, Adriana, Damián Ruiz-Ramoni, Paolo Ramoni-Perazzi, and Mariana Muñoz-Romo. 2019. Bat Folivory in Numbers: How Many, How Much, and How Long? *Acta Chiropterologica* 21(1):183–191.

Fernandez, Ana Z., Alfonso Tablante, Suzette Beguín, H. Coenraad Hemker, and Rafael Apitz-Castro. 1999. Draculin, the Anticoagulant Factor in Vampire Bat Saliva, Is a Tight-Binding, Noncompetitive Inhibitor of Activated Factor X. *Biochimica et Biophysica Acta: Protein Structure and Molecular Enzymology* 1434(1):135–142.

Gonzalez-Terrazas, Tania P., Rodrigo A. Medellín, Mirjam Knörnschild, and Marco Tschapka. 2012. Morphological Specialization Influences Nectar Extraction

Efficiency of Sympatric Nectar-Feeding Bats. *Journal of Experimental Biology* 215(22):3989–3996.

Gual-Suárez, Fernando, and Rodrigo A. Medellín. 2021. We Eat Meat: A Review of Carnivory in Bats. *Mammal Review* 51(4):540–558.

Harper, Cally J., Sharon M. Swartz, and Elizabeth L. Brainerd. 2013. Specialized Bat Tongue Is a Hemodynamic Nectar Mop. *Proceedings of the National Academy of Sciences of the United States of America* 110(22):8852–8857.

Holderied, Marc, Carmi Korine, and Thorsten Moritz. 2011. Hemprich's Long-Eared Bat (*Otonycteris hemprichii*) as a Predator of Scorpions: Whispering Echo-location, Passive Gleaning and Prey Selection. *Journal of Comparative Physiology A* 197(5):425–433.

Hopp, Bradley H., Ryan S. Arvidson, Michael E. Adams, and Khaleel A. Razak. 2017. Arizona Bark Scorpion Venom Resistance in the Pallid Bat, *Antrozous pallidus*. *PLoS ONE* 12(8):e0183215.

Howell, D. J., and Norman Hodgkin. 1976. Feeding Adaptations in the Hairs and Tongues of Nectar-Feeding Bats. *Journal of Morphology* 148(3):329–336.

Ito, Fernanda, Enrico Bernard, and Rodrigo A. Torres. 2016. What Is for Dinner? First Report of Human Blood in the Diet of the Hairy-Legged Vampire Bat *Diphylla ecaudata*. *Acta Chiropterologica* 18(2):509–515.

Kalka, Margareta, and Elisabeth K. V. Kalko. 2006. Gleaning Bats as Underestimated Predators of Herbivorous Insects: Diet of *Micronycteris microtis* (Phyllostomidae) in Panama. *Journal of Tropical Ecology* 22:1–10.

Muchhala, Nathan. 2006. Nectar Bat Stows Huge Tongue in Its Rib Cage. *Nature* 444(7120):701–702.

Neil, Thomas R., Zhiyuan Shen, Daniel Robert, Bruce W. Drinkwater, and Marc W. Holderied. 2020. Thoracic Scales of Moths as a Stealth Coating Against Bat Biosonar. *Journal of the Royal Society, Interface* 17(163).

Nelson, Suzanne L., Thomas H. Kunz, and Stephen R. Humphrey. 2005. Folivory in Fruit Bats: Leaves Provide a Natural Source of Calcium. *Journal of Chemical Ecology* 31(8):1683–1691.

Nogueira, Marcelo R., and Adriano L Peracchi. 2003. Fig-Seed Predation by 2 Species of *Chiroderma*: Discovery of New Feeding Strategy in Bats. *Journal of Mammalogy* 84(1):225–233.

Nogueira, Marcelo R., Adriano L. Peracchi, and Leandro R. Monteiro. 2009. Morphological Correlates of Bite Force and Diet in the Skull and Mandible of Phyllostomid Bats. *Functional Ecology* 23(4):715–723.

Otálora-Ardila, Aída, L. Gerardo Herrera, José Juan Flores-Martínez, and Christian C. Voigt. 2013. Marine and Terrestrial Food Sources in the Diet of the Fish-Eating Myotis (*Myotis vivesi*). *Journal of Mammalogy* 94(5):1102–1110.

Popa-Lisseanu, Ana G., Antonio Delgado-Huertas, Manuela G. Forero, Alicia Rodríguez, Raphaël Arlettaz, and Carlos Ibáñez. 2007. Bats' Conquest of a Formidable Foraging Niche: The Myriads of Nocturnally Migrating Songbirds. *PLoS ONE* 2(2):e205.

Santana, Sharlene E., and Elena Cheung. 2016. Go Big or Go Fish: Morphological Specializations in Carnivorous Bats. *Proceedings of the Royal Society B: Biological Sciences* 283:20160615.

Santana, Sharlene E., Elizabeth R. Dumont, and Julian L. Davis. 2010. Mechanics of Bite Force Production and Its Relationship to Diet in Bats. *Functional Ecology* 24(4):776–784.

Schleuning, Wolf Dieter. 2002. Vampire Bat Plasminogen Activator DSPA-Alpha-1 (Desmoteplase): A Thrombolytic Drug Optimized by Natural Selection. *Pathophysiology of Haemostasis and Thrombosis* 31(3–6):118–122.

Start, A. N., N. L. McKenzie, and R. D. Bullen. 2019. Notes on Bats in the Diets of Ghost Bats (*Macroderma gigas*: Megadermatidae) in the Pilbara Region of Western Australia. *Records of the Western Australian Museum* 34(1):51–53.

Swift, S. M., and P. A. Racey. 2002. Gleaning as a Foraging Strategy in Natterer's Bat *Myotis nattereri*. *Behavioral Ecology and Sociobiology* 52(5):408–416.

Tschapka, Marco, Tania P. Gonzalez-Terrazas, and Mirjam Knörnschild. 2015. Nectar Uptake in Bats Using a Pumping-Tongue Mechanism. *Science Advances* 1(8):1500525.

Vehrencamp, Sandra L., F. Gary Stiles, and Jack W. Bradbury. 1977. Observations on the Foraging Behavior and Avian Prey of the Neotropical Carnivorous Bat, *Vampyrum spectrum*. *Journal of Mammalogy* 58(4):469–478.

Winter, York, and Otto Von Helversen. 2003. Operational Tongue Length in Phyllostomid Nectar-Feeding Bats. *Journal of Mammalogy* 84(3):886–896.

Zepeda Mendoza, M. Lisandra, Zijun Xiong, Marina Escalera-Zamudio, Anne Kathrine Runge, Julien Thézé, Daniel Streicker, Hannah K. Frank, et al. 2018. Hologenomic Adaptations Underlying the Evolution of Sanguivory in the Common Vampire Bat. *Nature Ecology & Evolution* 2(4):659–668.

# 6. Flight

Amador, Lucila I., Nancy B. Simmons, and Norberto P. Giannini. 2019. Aerodynamic Reconstruction of the Primitive Fossil Bat *Onychonycteris finneyi* (Mammalia: Chiroptera). *Biology Letters* 15(3).

Brown, Emily E., Daniel D. Cashmore, Nancy B. Simmons, and Richard J. Butler. 2019. Quantifying the completeness of the bat fossil record. *Palaeontology* 62(5):757–776.

Cooper, Lisa Noelle, and Karen E. Sears. 2013. How to Grow a Bat Wing. In *Bat Evolution, Ecology, and Conservation*. Edited by Rick A. Adams and Scott C. Pedersen. New York: Springer. 3–20.

Farney, John, and Eugene D. Fleharty. 1969. Aspect Ratio, Loading, Wing Span, and Membrane Areas of Bats. *Journal of Mammalogy* 50(2):362.

Findley, J. S., E. H. Studier, and D. E. Wilson. 1972. Morphologic Properties of Bat Wings. *Journal of Mammalogy* 53(3):429–444.

Hand, Suzanne J., Vera Weisbecker, Robin M. D. Beck, Michael Archer, Henk Godthelp, Alan J. D. Tennyson, and Trevor H. Worthy. 2009. Bats That Walk: A New Evolutionary Hypothesis for the Terrestrial Behaviour of New Zealand's Endemic Mystacinids. *BMC Evolutionary Biology* 9(1):1–13.

Jones, Matthew F., and Stephen T. Hasiotis. 2018. Terrestrial Behavior and Trackway Morphology of Neotropical Bats. *Acta Chiropterologica* 20(1):229–250.

Marshall, Kara L., Mohit Chadha, Laura A. deSouza, Susanne J. Sterbing-D'Angelo, Cynthia F. Moss, and Ellen A. Lumpkin. 2015. Somatosensory Substrates of Flight Control in Bats. *Cell Reports* 11(6):851–858.

Riskin, Daniel K., and John W. Hermanson. 2005. Independent Evolution of Running in Vampire Bats. *Nature* 434(7031):292.

Riskin, Dan, Stuart Parsons, William A Schutt Jr., Gerald G. Carter, and John W. Hermanson. 2006. Terrestrial Locomotion of the New Zealand Short-Tailed Bat *Mystacina tuberculata* and the Common Vampire Bat *Desmodus rotundus*. *Journal of Experimental Biology* 209(9):1725–1736.

Rummel, Andrea D., Melissa M. Sierra, Brooke L. Quinn, and Sharon M. Swartz. 2023. Hair, There and Everywhere: A Comparison of Bat Wing Sensory Hair Distribution. *The Anatomical Record* 306(11):2681–2692.

Sears, Karen E., Richard R. Behringer, John J. Rasweiler IV, and Lee A. Niswander. 2006. Development of Bat Flight: Morphologic and Molecular Evolution of

Bat Wing Digits. *Proceedings of the National Academy of Sciences of the United States of America* 103(17):6581–6586.

Simmons, Nancy B., Kevin L. Seymour, Jörg Habersetzer, and Gregg F. Gunnell. 2008. Primitive Early Eocene Bat from Wyoming and the Evolution of Flight and Echolocation. *Nature* 451(7180):818–821.

Simmons, Nancy B., Kevin L. Seymour, Jörg Habersetzer, and Gregg F. Gunnell. 2010. Inferring Echolocation in Ancient Bats. *Nature* 466(7309):E8.

Springer, Mark S., Emma C. Teeling, Ole Madsen, Michael J. Stanhope, and Wilfried W. de Jong. 2001. Integrated Fossil and Molecular Data Reconstruct Bat Echolocation. *Proceedings of the National Academy of Sciences of the United States of America* 98(11):6241–6246.

Sterbing-D'Angelo, Susanne, Mohit Chadha, Chen Chiu, Ben Falk, Wei Xian, Janna Barcelo, John M. Zook, and Cynthia F. Moss. 2011. Bat Wing Sensors Support Flight Control. *Proceedings of the National Academy of Sciences of the United States of America* 108(27):11291–11296.

Sterbing-D'Angelo, S. J., M. Chadha, K. L. Marshall, and C. F. Moss. 2017. Functional Role of Airflow-Sensing Hairs on the Bat Wing. *Journal of Neurophysiology* 117(2):705–712.

Swartz, S. M., M. S. Groves, H. D. Kim, and W. R. Walsh. 1996. Mechanical Properties of Bat Wing Membrane Skin. *Journal of Zoology* 239(2):357–378.

Veselka, Nina, David D. McErlain, David W. Holdsworth, Judith L. Eger, Rethy K. Chhem, Matthew J. Mason, Kirsty L. Brain, Paul A. Faure, and M. Brock Fenton. 2010. A Bony Connection Signals Laryngeal Echolocation in Bats. *Nature* 463(7283):939–942.

# 7. Share

Adams, Amanda M., Kaylee Davis, and Michael Smotherman. 2017. Suppression of Emission Rates Improves Sonar Performance by Flying Bats. *Scientific Reports* 7(1):1–9.

Amichai, Eran, Gaddi Blumrosen, and Yossi Yovel. 2015. Calling Louder and Longer: How Bats Use Biosonar Under Severe Acoustic Interference from Other Bats. *Proceedings of the Royal Society B: Biological Sciences* 282(1821):20152064.

Arnold, Bryan D., and Gerald S. Wilkinson. 2011. Individual Specific Contact Calls of Pallid Bats (*Antrozous pallidus*) Attract Conspecifics at Roosting Sites. *Behavioral Ecology and Sociobiology* 65(8):1581–1593.

Barlow, Kate E., and Gareth Jones. 1997. Function of Pipistrelle Social Calls: Field Data and a Playback Experiment. *Animal Behaviour* 53:5190.

Beleyur, Thejasvi, and Holger R. Goerlitz. 2019. Modeling Active Sensing Reveals Echo Detection Even in Large Groups of Bats. *Proceedings of the National Academy of Sciences of the United States of America* 116(52):26662–26668.

Carter, Gerald G., and Gerald S. Wilkinson. 2013. Food Sharing in Vampire Bats: Reciprocal Help Predicts Donations More than Relatedness or Harassment. *Proceedings of the Royal Society B: Biological Sciences* 280(1753):20122573.

Carter, Gerald G., and Gerald S. Wilkinson. 2016. Common Vampire Bat Contact Calls Attract Past Food-Sharing Partners. *Animal Behaviour* 116:45–51.

Carter, Gerald, Diana Schoeppler, Marie Manthey, Mirjam Knörnschild, and Annette Denzinger. 2015. Distress Calls of a Fast-Flying Bat (*Molossus molossus*) Provoke Inspection Flights But Not Cooperative Mobbing. *PLoS ONE* 10(9):e0136146.

Carter, Gerald G., Damien R. Farine, Rachel J. Crisp, Julia K. Vrtilek, Simon P. Ripperger, and Rachel A. Page. 2020. Development of New Food-Sharing Relationships in Vampire Bats. *Current Biology* 30(7):1275–1279.e3.

Chiu, Chen, Wei Xian, and Cynthia F. Moss. 2008. Flying in Silence: Echolocating Bats Cease Vocalizing to Avoid Sonar Jamming. *Proceedings of the National Academy of Sciences of the United States of America* 105(35):13116–13121.

Corcoran, Aaron J., and William E. Conner. 2014. Bats Jamming Bats: Food Competition Through Sonar Interference. *Science* 346(6210):745–747.

Cvikel, Noam, Katya Egert Berg, Eran Levin, Edward Hurme, Ivailo Borissov, Arjan Boonman, Eran Amichai, and Yossi Yovel. 2015. Bats Aggregate to Improve Prey Search But Might Be Impaired When Their Density Becomes Too High. *Current Biology* 25(2):206–211.

Eckenweber, Maria, and Mirjam Knörnschild. 2016. Responsiveness to Conspecific Distress Calls Is Influenced by Day-Roost Proximity in Bats (*Saccopteryx bilineata*). *Royal Society Open Science* 3(5).

Egert-Berg, Katya, Edward R. Hurme, Stefan Greif, Aya Goldstein, Lee Harten, Luis Gerardo Herrera M., José Juan Flores-Martínez, et al. 2018. Resource Ephemerality Drives Social Foraging in Bats. *Current Biology* 28(22):3667–3673.e5.

Gillam, Erin H., Nachum Ulanovsky, and Gary F. McCracken. 2006. Rapid Jamming Avoidance in Biosonar. *Proceedings of the Royal Society B: Biological Sciences* 274(1610):651–660.

Hase, Kazuma, Yukimi Kadoya, Yosuke Maitani, Takara Miyamoto, Kohta I. Kobayasi, and Shizuko Hiryu. 2018. Bats Enhance Their Call Identities to Solve the Cocktail Party Problem. *Communications Biology* 1(1):1–8.

Jarvis, Jenna, William Jackson, and Michael Smotherman. 2013. Groups of Bats Improve Sonar Efficiency Through Mutual Suppression of Pulse Emissions. *Frontiers in Physiology* 4:140.

Lin, Yuan, Nicole Abaid, and Rolf Müller. 2016. Bats Adjust Their Pulse Emission Rates with Swarm Size in the Field. *Journal of the Acoustical Society of America* 140(6):4318.

Mazar, Omer, and Yossi Yovel. 2020. A Sensorimotor Model Shows Why a Spectral Jamming Avoidance Response Does Not Help Bats Deal with Jamming. *eLife* 9:e55539.

O'Mara, M. Teague, and Dina K. N. Dechmann. 2023. Greater Spear-Nosed Bats in Panama Do Not Use Social Proximity to Improve Foraging Efficiency. *bioRxiv* doi.org/10.1101/2021.09.30.462631.

Ripperger, Simon P., and Gerald G. Carter. 2021. Social Foraging in Vampire Bats Is Predicted by Long-Term Cooperative Relationships. *PLoS Biology* 19(9):e3001366.

Ripperger, Simon P., Gerald G. Carter, Niklas Duda, Alexander Koelpin, Björn Cassens, Rüdiger Kapitza, Darija Josic, Jineth Berrío-Martínez, Rachel A. Page, and Frieder Mayer. 2019. Vampire Bats That Cooperate in the Lab Maintain Their Social Networks in the Wild. *Current Biology* 29(23):4139–4144.

Ulanovsky, Nachum, M. Brock Fenton, Asaf Tsoar, and Carmi Korine. 2004. Dynamics of Jamming Avoidance in Echolocating Bats. *Proceedings of the Royal Society B: Biological Sciences* 271(1547):1467–1475.

Wilkinson, Gerald S. 1990. Food Sharing in Vampire Bats. *Scientific American* 262(2):76–83.

Wilkinson, Gerald S., and Janette Wenrick Boughman. 1998. Social Calls Coordinate Foraging in Greater Spear-Nosed Bats. *Animal Behaviour* 55:337–350.

# 8. Home

Avila-Flores, Rafael, and Rodrigo A. Medellín. 2004. Ecological, Taxonomic, and Physiological Correlates of Cave Use by Mexican Bats. *Journal of Mammalogy* 85(4):675–687.

Balasingh, J., John Koilraj, and Thomas H. Kunz. 1995. Tent Construction by the Short-Nosed Fruit Bat *Cynopterus sphinx* (Chiroptera: Pteropodidae) in Southern India. *Ethology* 100(3):210–229.

Choe, J. C. 2007. Ingenious Design of Tent Roosts by Peters's Tent-Making Bat, *Uroderma bilobatum* (Chiroptera: Phyllostomidae). *Journal of Natural History* 28(3):731–737.

Davis, Wayne H. 1970. Hibernation: Ecology and Physiological Ecology. In *Biology of Bats*. Edited by William A. Wimsatt. Philadelphia, PA: Elsevier Science and Technology Books. 265–300.

Dechmann, Dina K. N., Sharlene E. Santana, and Elizabeth R. Dumont. 2009. Roost Making in Bats: Adaptations for Excavating Active Termite Nests. *Journal of Mammalogy* 90(6):1461–1468.

Dorward, Leejiah J., Adam Stein, and Charlotte Searle. 2022. Like a Bat out of Hell: Bats Roosting in Pit-Latrine Cesspits. *African Journal of Ecology* 60(1):91–94.

Dunbar, Miranda B., and Thomas E. Tomasi. 2006. Arousal Patterns, Metabolic Rate, and an Energy Budget of Eastern Red Bats (*Lasiurus borealis*) in Winter. *Journal of Mammalogy* 87(6):1096–1102.

Fleming, Theodore H., Peggy Eby, T. H. Kunz, and M. Brock Fenton. 2003. Ecology of Bat Migration. In *Bat Ecology*. Edited by Thomas H. Kunz and M. Brock Fenton. Chicago: University of Chicago Press. 156–165.

Furey, Neil M., and Paul A. Racey. 2015. Conservation Ecology of Cave Bats. In *Bats in the Anthropocene: Conservation of Bats in a Changing World*. Edited by Christian C. Voigt and Tigga Kingston. New York: Springer. 463–500.

Galván, Ismael, Juan Carlos Vargas-Mena, and Bernal Rodríguez-Herrera. 2020. Tent-Roosting May Have Driven the Evolution of Yellow Skin Coloration in Stenodermatinae Bats. *Journal of Zoological Systematics and Evolutionary Research* 58(1):519–527.

Hirakawa, Hirofumi, and Yu Nagasaka. 2018. Evidence for Ussurian Tube-Nosed Bats (*Murina ussuriensis*) Hibernating in Snow. *Scientific Reports* 8(1):1–8.

Howell, D. J., and Joe Pylka. 1977. Why Bats Hang Upside Down: A Biomechanical Hypothesis. *Journal of Theoretical Biology* 69(4):625–631.

Hurme, Edward, Jakob Fahr, Bakwo Fils Eric-Moise, C. Tom Hash, M. Teague O'Mara, Heidi Richter, Iroro Tanshi, et al. 2022. Fruit Bat Migration Matches Green Wave in Seasonal Landscapes. *Functional Ecology* 36(8):2043–2055.

Kalko, Elisabeth K. V., Katja Ueberschaer, and Dina Dechmann. 2006. Roost Structure, Modification, and Availability in the White-Throated Round-Eared Bat,

*Lophostoma silvicolum* (Phyllostomidae) Living in Active Termite Nests. *Biotropica* 38(3):398–404.

Kunz, Thomas H., Linda F. Lumsden, and M. Brock Fenton. 2003. Ecology of Cavity and Foliage Roosting Bats. In *Bat Ecology*. Edited by Thomas H. Kunz and M. Brock Fenton. Chicago: University of Chicago Press. 3–89.

Lehnert, Linn S., Stephanie Kramer-Schadt, Tobias Teige, Uwe Hoffmeister, Ana Popa-Lisseanu, Fabio Bontadina, Mateusz Ciechanowski, et al. 2018. Variability and Repeatability of Noctule Bat Migration in Central Europe: Evidence for Partial and Differential Migration. *Proceedings of the Royal Society B* 285(1893):20182174.

Lewis, S. E. 1995. Roost Fidelity of Bats: A Review. *Journal of Mammalogy* 76(2):481–496.

McClure, Meredith L., Daniel Crowley, Catherine G. Haase, Liam P. McGuire, Nathan W. Fuller, David T. S. Hayman, Cori L. Lausen, Raina K. Plowright, Brett G. Dickson, and Sarah H. Olson. 2020. Linking Surface and Subterranean Climate: Implications for the Study of Hibernating Bats and Other Cave Dwellers. *Ecosphere* 11(10):e03274.

Medway, Lord, and Adrian G. Marshall. 1970. Roost-Site Selection Among Flat-Headed Bats (*Tylonycteris* spp.). *Journal of Zoology* 161:237–245.

Mel, R. K. de, A. P. Sumanapala, H. D. Jayasinghe, S. S. Rajapakshe, and R. P. Nanayakkara. 2021. An Unusual Roosting Habit of a Painted Bat (*Kerivoula picta*) from Sri Lanka. *Taprobanica* 10(2):138–139.

Mormann, Brad M., and Lynn W. Robbins. 2007. Winter Roosting Ecology of Eastern Red Bats in Southwest Missouri. *Journal of Wildlife Management* 71(1):213–217.

Morrison, Douglas W. 1980. Foraging and Day-Roosting Dynamics of Canopy Fruit Bats in Panama. *Journal of Mammalogy* 61(1):20–29.

O'Mara, M. Teague, Dina K. N. Dechmann, and Rachel A. Page. 2014. Frugivorous Bats Evaluate the Quality of Social Information When Choosing Novel Foods. *Behavioral Ecology* 25(5):1233–1239.

Parker-Shames, Phoebe, Bernal Rodriguez Herrera, Valeria Da, and Cunha Tavares. 2013. Maximum Weight Capacity of Leaves Used by Tent-Roosting Bats: Implications for Social Structure. *Chiroptera Neotropical* 19(3):36–43.

Perry, Roger W. 2013. A Review of Factors Affecting Cave Climates for Hibernating Bats in Temperate North America. *Environmental Reviews* 21(1):28–39.

Quinn, Thomas H., and Julian J. Baumel. 1993. Chiropteran Tendon Locking Mechanism. *Journal of Morphology* 216(2):197–208.

Ramakers, Jip J. C., Dina K. N. Dechmann, Rachel A. Page, and M. Teague O'Mara. 2016. Frugivorous Bats Prefer Information from Novel Social Partners. *Animal Behaviour* 116:83–87.

Richter, H. V., and G. S. Cumming. 2006. Food Availability and Annual Migration of the Straw-Colored Fruit Bat (*Eidolon helvum*). *Journal of Zoology* 268(1):35–44.

Riskin, Daniel K., and M. Brock Fenton. 2001. Sticking Ability in Spix's Disk-Winged Bat, *Thyroptera tricolor* (Microchiroptera: Thyropteridae). *Canadian Journal of Zoology* 79(12):2261–2267.

Riskin, Daniel K., and Paul A. Racey. 2010. How Do Sucker-Footed Bats Hold On and Why Do They Roost Head-Up? *Biological Journal of the Linnean Society* 99(2):233–240.

Rojas-Martínez, Alberto, Alfonso Valiente-Banuet, Maria Del Coro Arizmendi, Ariel Alcántara-Eguren, and Héctor T. Arita. 1999. Seasonal Distribution of the Long-Nosed Bat (*Leptonycteris curasoae*) in North America: Does a Generalized Migration Pattern Really Exist? *Journal of Biogeography* 26(5):1065–1077.

Sagot, Maria, and Richard D. Stevens. 2012. The Evolution of Group Stability and Roost Lifespan: Perspectives from Tent-Roosting Bats. *Biotropica* 44(1):90–97.

Santana, Sharlene E., Thomas O. Dial, Thomas P. Eiting, and Michael E. Alfaro. 2011. Roosting Ecology and the Evolution of Pelage Markings in Bats. *PLoS ONE* 6(10):e25845.

Schöner, Michael G., Caroline R. Schöner, Gerald Kerth, Siti Nurqayah Binti Pg Suhaini, and T. Ulmar Grafe. 2017. Handle with Care: Enlarged Pads Improve the Ability of Hardwicke's Woolly Bat, *Kerivoula hardwickii* (Chiroptera: Vespertilionidae), to Roost in a Carnivorous Pitcher Plant. *Biological Journal of the Linnean Society* 122(3):643–650.

Schulz, M. 2000. Roosts Used by the Golden-Tipped Bat *Kerivoula papuensis* (Chiroptera: Vespertilionidae). *Journal of Zoology* 250(4):467–478.

Vasenkov, Denis, Jean François Desmet, Igor Popov, and Natalia Sidorchuk. 2022. Bats Can Migrate Farther than It Was Previously Known: A New Longest Migration Record by Nathusius' Pipistrelle *Pipistrellus nathusii* (Chiroptera: Vespertilionidae). *Mammalia* 86(5):524–526.

Webala, Paul W., Simon Musila, Robert Syingi, and Zedekiah A. Okwany. 2022. Bats in Kenyan Pit Latrines: Non-Invasive Sampling by Photography. *African Journal of Ecology* 60(3):834–837.

Weller, Theodore J., Kevin T. Castle, Felix Liechti, Cris D. Hein, Michael R. Schirmacher, and Paul M. Cryan. 2016. First Direct Evidence of Long-Distance Seasonal Movements and Hibernation in a Migratory Bat. *Scientific Reports* 6(1):1–7.

# 9. Love

Adams, Danielle M., Yue Li, and Gerald S. Wilkinson. 2018. Male Scent Gland Signals Mating Status in Greater Spear-Nosed Bats, *Phyllostomus hastatus*. *Journal of Chemical Ecology* 44(11):975–986.

Behr, Oliver, and Otto Von Helversen. 2004. Bat Serenades: Complex Courtship Songs of the Sac-Winged Bat (*Saccopteryx bilineata*). *Behavioral Ecology and Sociobiology* 56(2):106–115.

Bohn, Kirsten M., Barbara Schmidt-French, Christine Schwartz, Michael Smotherman, and George D. Pollak. 2009. Versatility and Stereotypy of Free-Tailed Bat Songs. *PLoS ONE* 4(8):e6746.

Bradbury, Jack W. 1977. Lek Mating Behavior in the Hammer-Headed Bat. *Zeitschrift Für Tierpsychologie* 45(3):225–255.

Caspers, Barbara, Stephan Franke, and Christian C. Voigt. 2008. The Wing-Sac Odour of Male Greater Sac-Winged Bats *Saccopteryx bilineata* (Emballonuridae) as a Composite Trait: Seasonal and Individual Differences. In *Chemical Signals in Vertebrates 11*. Edited by Jane L. Hurst, Robert J. Beynon, S. Craig Roberts, Tristram D. Wyatt. New York: Springer. 151–160.

Caspers, Barbara A., Frank C. Schroeder, Stephan Franke, W. Jürgen Streich, and Christian C. Voigt. 2009. Odour-Based Species Recognition in Two Sympatric Species of Sac-Winged Bats (*Saccopteryx bilineata, S. leptura*): Combining Chemical Analyses, Behavioural Observations and Odour Preference Tests. *Behavioral Ecology and Sociobiology* 63(5):741–749.

Collier, Kathleen, and Stuart Parsons. 2022. Syntactic Properties of Male Courtship Song in the Lesser Short-Tailed Bat, *Mystacina tuberculata*. *Frontiers in Ecology and Evolution* 10:907791.

Cryan, Paul M., Joel W. Jameson, Erin F. Baerwald, Craig K. R. Willis, Robert M. R. Barclay, E. Apple Snider, and Elizabeth G. Crichton. 2012. Evidence of Late-Summer Mating Readiness and Early Sexual Maturation in Migratory Tree-Roosting Bats Found Dead at Wind Turbines. *PLoS ONE* 7(10):e47586.

Fasel, Nicolas J., Mnqobi L. Mamba, and Ara Monadjem. 2020. Penis Morphology Facilitates Identification of Cryptic African Bat Species. *Journal of Mammalogy* 101(5):1392–1399.

Faulkes, Chris G., J. Stephen Elmore, David A. Baines, M. Brock Fenton, Nancy B. Simmons, and Elizabeth L. Clare. 2019. Chemical Characterisation of Potential Pheromones from the Shoulder Gland of the Northern Yellow-Shouldered Bat, *Sturnira parvidens* (Phyllostomidae: Stenodermatinae). *Peer Journal* 2019(9):e7734.

Flores, Victoria, and Rachel A. Page. 2017. Novel Odorous Crust on the Forearm of Reproductive Male Fringe-Lipped Bats (*Trachops cirrhosus*). *Journal of Mammalogy* 98(6):1568–1577.

Flores, Victoria, Jill M. Mateo, and Rachel A. Page. 2019. The Role of Male Forearm Crust Odour in Fringe-Lipped Bats (*Trachops cirrhosus*). *Behaviour* 156(15):1435–1458.

Harten, Lee, Yosef Prat, Shachar Ben Cohen, Roi Dor, and Yossi Yovel. 2019. Food for Sex in Bats Revealed as Producer Males Reproduce with Scrounging Females. *Current Biology* 29(11):1895–1900.e3.

Heideman, P. D., K. R. Erickson, and J. B. Bowles. 1990. Notes on the Breeding Biology, Gular Gland and Roost Habits of *Molossus sinaloe* (Chiroptera, Molossidae). *Zeitschrift fuer Saeugetierkunde* 55:303–307.

Maruthupandian, Jayabalan, and Ganapathy Marimuthu. 2013. Cunnilingus Apparently Increases Duration of Copulation in the Indian Flying Fox, *Pteropus giganteus*. *PLoS ONE* 8(3):e59743.

McCracken, Gary F., and Gerald S. Wilkinson. 2000. Bat Mating Systems. In *Reproductive Biology of Bats*. Edited by Elizabeth G. Crichton and Philip H. Krutzsch. Philadelphia, PA: Elsevier. 321–362.

Muñoz-Romo, Mariana, Juan F. Burgos, and Thomas H. Kunz. 2011. Smearing Behaviour of Male *Leptonycteris curasoae* (Chiroptera) and Female Responses to the Odour of Dorsal Patches. *Behaviour* 148(4):461–483.

Muñoz-Romo, Mariana, Lawrence T. Nielsen, Jafet M. Nassar, and Thomas H. Kunz. 2012. Chemical Composition of the Substances from Dorsal Patches of Males of the Curaçaoan Long-Nosed Bat, *Leptonycteris curasoae* (Phyllostomidae: Glossophaginae). *Acta Chiropterologica* 14(1):213–224.

Muñoz-Romo, Mariana, Rachel A. Page, and Thomas H. Kunz. 2021. Redefining the Study of Sexual Dimorphism in Bats: Following the Odour Trail. *Mammal Review* 51(2):155–177.

Ober, Holly K., Elizabeth C. Braun De Torrez, Jeffery A. Gore, Amanda M. Bailey, Jennifer K. Myers, Kathleen N. Smith, and Robert A. McCleery. 2017. Social Organization of an Endangered Subtropical Species, *Eumops floridanus*, the Florida Bonneted Bat. *Mammalia* 81(4):375–383.

Orr, Teri J., Theresa Lukitsch, Thomas P. Eiting, and Patricia L. R. Brennan. 2022. Testing Morphological Relationships Between Female and Male Copulatory Structures in Bats. *Integrative and Comparative Biology* 62(3):602–612.

Rodríguez-Herrera, Bernal, Ricardo Sánchez-Calderón, Victor Madrigal-Elizondo, Paulina Rodríguez, Jairo Villalobos, Esteban Hernández, Daniel Zamora-Mejías, Gloria Gessinger, and Marco Tschapka. 2020. The Masked Seducers: Lek Courtship Behavior in the Wrinkle-Faced Bat *Centurio senex* (Phyllostomidae). *PLoS ONE* 15(11):e0241063.

Russ, J. M., and P. A. Racey. 2007. Species-Specificity and Individual Variation in the Song of Male Nathusius' Pipistrelles (*Pipistrellus nathusii*). *Behavioral Ecology and Sociobiology* 61(5):669–677.

Smotherman, Michael, Mirjam Knörnschild, Grace Smarsh, and Kirsten Bohn. 2016. The Origins and Diversity of Bat Songs. *Journal of Comparative Physiology A* 202(8):535–554.

Tan, Min, Gareth Jones, Guangjian Zhu, Jianping Ye, Tiyu Hong, Shanyi Zhou, Shuyi Zhang, and Libiao Zhang. 2009. Fellatio by Fruit Bats Prolongs Copulation Time. *PLoS ONE* 4(10):e7595.

Toth, Cory A., and Stuart Parsons. 2013. Is Lek Breeding Rare in Bats? *Journal of Zoology* 291(1):3–11.

Toth, Cory A., and Stuart Parsons. 2018. The High-Output Singing Displays of a Lekking Bat Encode Information on Body Size and Individual Identity. *Behavioral Ecology and Sociobiology* 72(7):1–14.

Voigt, Christian C. 2002. Individual Variation in Perfume Blending in Male Greater Sac-Winged Bats. *Animal Behaviour* 63(5):907–913.

Voigt, Christian C., Oliver Behr, Barbara Caspers, Otto Von Helversen, Mirjam Knörnschild, Frieder Mayer, and Martina Nagy. 2008. Songs, Scents, and Senses: Sexual Selection in the Greater Sac-Winged Bat, *Saccopteryx bilineata*. *Journal of Mammalogy* 89(6):1401–1410.

# 10. Grow

Adams, Rick, and Jason Shaw. 2013. Time's Arrow in the Evolutionary Development of Bat Flight. In *Bat Evolution, Ecology, and Conservation*. Edited by Rick A. Adams and Scott C. Pedersen. New York: Springer. 21–46.

Badwaik, Nilima K., and John J. Rasweiler. 2000. Pregnancy. In *Reproductive Biology of Bats*. Edited by Elizabeth G. Crichton and Philip H. Krutzsch. Philadelphia, PA: Elsevier. 221–293.

Bernard, R. T. F., and G. S. Gumming. 1997. African Bats: Evolution of Reproductive Patterns and Delays. *Quarterly Review of Biology* 72(3):253–274.

Bohn, Kirsten M., Cynthia F. Moss, and Gerald S. Wilkinson. 2009. Pup Guarding by Greater Spear-Nosed Bats. *Behavioral Ecology and Sociobiology* 63(12):1693–1703.

Brown, Patricia E., Alan D. Grinnell, and Jean B. Harrison. 1978. The Development of Hearing in the Pallid Bat, *Antrozous pallidus. Journal of Comparative Physiology A* 126(2):169–182.

Carter, Richard T., and Rick A. Adams. 2016. Integrating Ontogeny of Echolocation and Locomotion Gives Unique Insights into the Origin of Bats. *Journal of Mammalian Evolution* 23(4):413–421.

Carter, R. T., J. B. Shaw, and R. A. Adams. 2014. Ontogeny of Vocalization in Jamaican Fruit Bats with Implications for the Evolution of Echolocation. *Journal of Zoology* 293(1):25–32.

Crichton, Elizabeth G. 2000. Sperm Storage and Fertilization. In *Reproductive Biology of Bats*. Edited by Elizabeth G. Crichton and Philip H. Krutzsch. Philadelphia, PA: Elsevier. 295–320.

Francis, Charles M., Edythe L. P. Anthony, Jennifer A. Brunton, and Thomas H. Kunz. 1994. Lactation in Male Fruit Bats. *Nature* 367:691–692.

Geipel, Inga, Elisabeth K. V. Kalko, Katja Wallmeyer, and Mirjam Knörnschild. 2013. Postweaning Maternal Food Provisioning in a Bat with a Complex Hunting Strategy. *Animal Behaviour* 85(6):1435–1441.

Gelfand, Deborah L., and Gary F. McCracken. 1986. Individual Variation in the Isolation Calls of Mexican Free-Tailed Bat Pups (*Tadarida brasiliensis mexicana*). *Animal Behaviour* 34(4):1078–1086.

Goldshtein, Aya, Lee Harten, and Yossi Yovel. 2022. Mother Bats Facilitate Pup Navigation Learning. *Current Biology* 32(2):350–360.e4.

Gould, Edwin. 1975. Experimental Studies of the Ontogeny of Ultrasonic Vocalizations in Bats. *Developmental Psychobiology* 8(4):333–346.

Grunstra, Nicole D. S., Frank E. Zachos, Anna Nele Herdina, Barbara Fischer, Mihaela Pavličev, and Philipp Mitteroecker. 2019. Humans as Inverted Bats: A Comparative Approach to the Obstetric Conundrum. *American Journal of Human Biology* 31(2):e23227.

Gustin, Mary K., and Gary F. McCracken. 1987. Scent Recognition Between Females and Pups in the Bat *Tadarida brasiliensis mexicana*. *Animal Behaviour* 35(1):13–19.

Harten, Lee, Amitay Katz, Aya Goldshtein, Michal Handel, and Yossi Yovel. 2020. The Ontogeny of a Mammalian Cognitive Map in the Real World. *Science* 369(6500):194–197.

Hughes, P. M., J. M. V. Rayner, and G. Jones. 1995. Ontogeny of "True" Flight and Other Aspects of Growth in the Bat *Pipistrellus pipistrellus*. *Journal of Zoology* 236(2):291–318.

Knörnschild, Mirjam, Martina Nagy, Markus Metz, Frieder Mayer, and Otto Von Helversen. 2010. Complex Vocal Imitation During Ontogeny in a Bat. *Biology Letters* 6(2):156.

Kohles, Jenna E., Rachel A. Page, Dina K. N. Dechmann, and Teague O'Mara. 2018. Rapid Behavioral Changes During Early Development in Peters' Tent-Making Bat (*Uroderma bilobatum*). *PLoS ONE* 13(10):e0205351.

Kunz, Thomas H., and Wendy R. Hood. 2000. Parental Care and Postnatal Growth in the Chiroptera. In *Reproductive Biology of Bats*. Edited by Elizabeth G. Crichton and Philip H. Krutzsch. Philadelphia, PA: Elsevier. 415–468.

Kunz, Thomas H., and David J. Hosken. 2009. Male Lactation: Why, Why Not and Is It Care? *Trends in Ecology & Evolution* 24(2):80–85.

Kunz, T. H., A. L. Allgaier, K. Seyjagat, and R. Caligiuri. 1994. Allomaternal Care: Helper-Assisted Birth in the Rodrigues Fruit Bat, *Pteropus rodricensis* (Chiroptera: Pteropodidae). *Journal of Zoology* 232(4):691–700.

Kurta, A., G. P. Bell, K. A. Nagy, and T. H. Kunz. 1989. Energetics of Pregnancy and Lactation in Free-Ranging Little Brown Bats (*Myotis lucifugus*). *Physiological Zoology* 62(3):804–818.

Lewis, Susan E. 1992. Behavior of Peter's Tent-Making Bat, *Uroderma bilobatum*, at Maternity Roosts in Costa Rica. *Journal of Mammalogy* 73(3):541–546.

Loughry, W. J., and Gary F. McCracken. 1991. Factors Influencing Female-Pup Scent Recognition in Mexican Free-Tailed Bats. *Journal of Mammalogy* 72(3):624–626.

Mayberry, Heather W., and Paul A. Faure. 2015. Morphological, Olfactory, and Vocal Development in Big Brown Bats. *Biology Open* 4(1):22–34.

McCracken, Gary F. 1984. Communal Nursing in Mexican Free-Tailed Bat Maternity Colonies. *Science* 223(4640):1090–1091.

McCracken, Gary F. 1993. Locational Memory and Female-Pup Reunions in Mexican Free-Tailed Bat Maternity Colonies. *Animal Behaviour* 45(4):811–813.

McCracken, Gary F., and Mary K. Gustin. 1991. Nursing Behavior in Mexican Free-Tailed Bat Maternity Colonies. *Ethology* 89(4):305–321.

Navarro, Daniel L., and Don E. Wilson. 1982. *Vampyrum spectrum. Mammalian Species* 184:1–4.

Orr, Teri J., and Marlene Zuk. 2013. Does Delayed Fertilization Facilitate Sperm Competition in Bats? *Behavioral Ecology and Sociobiology* 67(12):1903–1913.

Pfeiffer, B., and F. Mayer. 2013. Spermatogenesis, Sperm Storage and Reproductive Timing in Bats. *Journal of Zoology* 289(2):77–85.

Racey, Paul A., and Abigail C. Entwistle. 2000. Life-History and Reproductive Strategies of Bats. In *Reproductive Biology of Bats*. Edited by Elizabeth G. Crichton and Philip H. Krutzsch. Philadelphia, PA: Elsevier. 363–414.

Razik, Imran, Bridget K. G. Brown, Rachel A. Page, and Gerald G. Carter. 2021. Non-Kin Adoption in the Common Vampire Bat. *Royal Society Open Science* 8:201927.

Reynolds, D. Scott, and Thomas H. Kunz. 2000. Changes in Body Composition During Reproduction and Postnatal Growth in the Little Brown Bat, *Myotis lucifugus* (Chiroptera: Vespertilionidae). *Ecoscience* 7(1):10–17.

Ripperger, Simon, Linus Günther, Hanna Wieser, Niklas Duda, Martin Hierold, Björn Cassens, Rüdiger Kapitza, Alexander Koelpin, and Frieder Mayer. 2019. Proximity Sensors on Common Noctule Bats Reveal Evidence That Mothers Guide Juveniles to Roosts But Not Food. *Biology Letters* 15(2):20180884.

Rose, Andreas, Marco Tschapka, and Mirjam Knörnschild. 2020. Visits at Artificial RFID Flowers Demonstrate That Juvenile Flower-Visiting Bats Perform Foraging Flights Apart from Their Mothers. *Mammalian Biology* 100(5):463–471.

Rydell, Jens. 1993. Variation in Foraging Activity of an Aerial Insectivorous Bat During Reproduction. *Journal of Mammalogy* 74(2):503–509.

Smarsh, Grace C., Yifat Tarnovsky, and Yossi Yovel. 2021. Hearing, Echolocation, and Beam Steering from Day 0 in Tongue-Clicking Bats. *Proceedings of the Royal Society B* 288(1961):20211714.

Wilkinson, Gerald S. 1992. Communal Nursing in the Evening Bat, *Nycticeius humeralis. Sociobiology* 31(4):225–235.

Wilkinson, Gerald S., Gerald G. Carter, Kirsten M. Bohn, and Danielle M. Adams. 2016. Non-Kin Cooperation in Bats. *Philosophical Transactions of the Royal Society B: Biological Sciences* 371(1687).

Williams, Lisa M., and Margaret C. Brittingham. 1997. Selection of Maternity Roosts by Big Brown Bats. *The Journal of Wildlife Management* 61(2):359–368.

Wilson, Don E. 2015. *Bats in Question: The Smithsonian Answer Book*. Washington, DC: Smithsonian Institution.

Wimsatt, William A. 1945. Notes on Breeding Behavior, Pregnancy, and Parturition in Some Vespertilionid Bats of the Eastern United States. *Journal of Mammalogy* 26(1):23–33.

# II. Spillover

Belotto, A., L. F. Leanes, M. C. Schneider, H. Tamayo, and E. Correa. 2005. Overview of Rabies in the Americas. *Virus Research* 111(1):5–12.

Beranová, Lucie, Marcin P. Joachimiak, Tomáš Kliegr, Gollam Rabby, and Vilém Sklenák. 2022. Why Was This Cited? Explainable Machine Learning Applied to COVID-19 Research Literature. *Scientometrics* 127(5):2313–2349.

Brokaw, Alyson F., Jeff Clerc, and Theodore James Weller. 2016. Another Account of Interspecific Aggression Involving a Hoary Bat (*Lasiurus cinereus*). *Northwestern Naturalist* 97(2):130–134.

Cerri, Jacopo, Emiliano Mori, Leonardo Ancillotto, Danilo Russo, and Sandro Bertolino. 2022. COVID-19, Media Coverage of Bats and Related Web Searches: A Turning Point for Bat Conservation? *Mammal Review* 52(1):16–25.

Cui, Jie, Fang Li, and Zheng Li Shi. 2018. Origin and Evolution of Pathogenic Coronaviruses. *Nature Reviews Microbiology* 17(3):181–192.

Davis, Patricia L., Hervé Bourhy, and Edward C. Holmes. 2006. The Evolutionary History and Dynamics of Bat Rabies Virus. *Infection, Genetics and Evolution* 6(6):464–473.

Drexler, Jan Felix, Victor Max Corman, Marcel Alexander Müller, Gael Darren Maganga, Peter Vallo, Tabea Binger, Florian Gloza-Rausch, et al. 2012. Bats Host Major Mammalian Paramyxoviruses. *Nature Communications* 3(1):1–13.

Epstein, Jonathan H., Simon J. Anthony, Ariful Islam, A. Marm Kilpatrick, Shahneaz Ali Khan, Maria D. Balkey, Noam Ross, et al. 2020. Nipah Virus Dynamics in Bats and Implications for Spillover to Humans. *Proceedings of the National Academy of Sciences of the United States of America* 117(46):29190–29201.

Fagre, Anna C., Lily E. Cohen, Evan A. Eskew, Max Farrell, Emma Glennon, Maxwell B. Joseph, Hannah K. Frank, Sadie J. Ryan, Colin J. Carlson, and Gregory F. Albery. 2022. Assessing the Risk of Human-to-Wildlife Pathogen Transmission for Conservation and Public Health. *Ecology Letters* 25(6):1534–1549.

Fairhead, James, Melissa Leach, and Dominique Millimouno. 2021. Spillover or Endemic? Reconsidering the Origins of Ebola Virus Disease Outbreaks by Revisiting Local Accounts in Light of New Evidence from Guinea. *BMJ Global Health* 6(4):e005783.

Fenton, M. Brock, Samira Mubareka, Susan M. Tsang, Nancy B. Simmons, and Daniel J. Becker. 2020. COVID-19 and Threats to Bats. *Facets* 5(1):349–352.

Fenton, M. Brock, Daniel G. Streicker, Paul A. Racey, Merlin D. Tuttle, Rodrigo A. Medellin, Mark J. Daley, Sergio Recuenco, and Kevin M. Bakker. 2020. Knowledge Gaps About Rabies Transmission from Vampire Bats to Humans. *Nature Ecology and Evolution* 4(4):517–518.

Field, Hume, Carol de Jong, Deb Melville, Craig Smith, Ina Smith, Alice Broos, Yu Hsin Nina Kung, Amanda McLaughlin, and Anne Zeddeman. 2011. Hendra Virus Infection Dynamics in Australian Fruit Bats. *PLoS ONE* 6(12):e28678.

Guth, Sarah, Nardus Mollentze, Katia Renault, Daniel G. Streicker, Elisa Visher, Mike Boots, and Cara E. Brook. 2022. Bats Host the Most Virulent—But Not the Most Dangerous—Zoonotic Viruses. *Proceedings of the National Academy of Sciences of the United States of America* 119(14):e2113628119.

Guy, Cylita, Jeneni Thiagavel, Nicole Mideo, and John M. Ratcliffe. 2019. Phylogeny Matters: Revisiting "A Comparison of Bats and Rodents as Reservoirs of Zoonotic Viruses." *Royal Society Open Science* 6(2):181182.

Guy, Cylita, John M. Ratcliffe, and Nicole Mideo. 2020. The Influence of Bat Ecology on Viral Diversity and Reservoir Status. *Ecology and Evolution* 10(12):5748–5758.

Han, Hui Ju, Hao Yu, and Xue Jie Yu. 2016. Evidence for Zoonotic Origins of Middle East Respiratory Syndrome Coronavirus. *The Journal of General Virology* 97(2):274.

Holmes, Edward C., Stephen A. Goldstein, Angela L. Rasmussen, David L. Robertson, Alexander Crits-Christoph, Joel O. Wertheim, Simon J. Anthony, et al. 2021. The Origins of SARS-CoV-2: A Critical Review. *Cell* 184(19):4848–4856.

Hu, Ben, Lei Ping Zeng, Xing Lou Yang, Xing Yi Ge, Wei Zhang, Bei Li, Jia Zheng Xie, et al. 2017. Discovery of a Rich Gene Pool of Bat SARS-Related Coronaviruses

Provides New Insights into the Origin of SARS Coronavirus. *PLoS Pathogens* 13(11):e1006698.

Ithete, Ndapewa Laudika, Samantha Stoffberg, Victor Max Corman, Veronika M. Cottontail, Leigh Rosanne Richards, M. Corrie Schoeman, Christian Drosten, Jan Felix Drexler, and Wolfgang Preiser. 2013. Close Relative of Human Middle East Respiratory Syndrome Coronavirus in Bat, South Africa. *Emerging Infectious Diseases* 19(10):1697–1699.

Jackson, Alan C., and M. Brock Fenton. 2001. Human Rabies and Bat Bites. *The Lancet* 357:1714.

Jones, Bryony A., Delia Grace, Richard Kock, Silvia Alonso, Jonathan Rushton, Mohammed Y. Said, Declan McKeever, et al. 2013. Zoonosis Emergence Linked to Agricultural Intensification and Environmental Change. *Proceedings of the National Academy of Sciences of the United States of America* 110(21):8399–8404.

Kuzmin, Ivan V., Brooke Bozick, Sarah A. Guagliardo, Rebekah Kunkel, Joshua R. Shak, Suxiang Tong, and Charles E. Rupprecht. 2011. Bats, Emerging Infectious Diseases, and the Rabies Paradigm Revisited. *Emerging Health Threats Journal* 4:1.

Lau, Susanna K. P., Patrick C. Y. Woo, Kenneth S. M. Li, Yi Huang, Hoi Wan Tsoi, Beatrice H. L. Wong, Samson S. Y. Wong, Suet Yi Leung, Kwok Hung Chan, and Kwok Yung Yuen. 2005. Severe Acute Respiratory Syndrome Coronavirus-Like Virus in Chinese Horseshoe Bats. *Proceedings of the National Academy of Sciences of the United States of America* 102(39):14040–14045.

Lau, Susanna K. P., Antonio C. P. Wong, Terrence C. K. Lau, and Patrick C. Y. Woo. 2017. Molecular Evolution of MERS Coronavirus: Dromedaries as a Recent Intermediate Host or Long-Time Animal Reservoir? *International Journal of Molecular Sciences* 18(10):2138.

Leroy, Eric M., Brice Kumulungui, Xavier Pourrut, Pierre Rouquet, Alexandre Hassanin, Philippe Yaba, André Délicat, Janusz T. Paweska, Jean Paul Gonzalez, and Robert Swanepoel. 2005. Fruit Bats as Reservoirs of Ebola Virus. *Nature* 438(7068):575–576.

Li, Wendong, Zhengli Shi, Meng Yu, Wuze Ren, Craig Smith, Jonathan H. Epstein, Hanzhong Wang, et al. 2005. Bats Are Natural Reservoirs of SARS-like Coronaviruses. *Science* 310(5748):676–679.

Lu, Manman, Xindong Wang, Huan Ye, Huimin Wang, Shan Qiu, Hongmao Zhang, Ying Liu, Jinhong Luo, and Jiang Feng. 2021. Does Public Fear That Bats Spread COVID-19 Jeopardize Bat Conservation? *Biological Conservation* 254:108952.

Luis, Angela D., David T. S. Hayman, Thomas J. O'Shea, Paul M. Cryan, Amy T. Gilbert, Angela D. Luis, David T. S. Hayman, et al. 2013. A Comparison of Bats and

Rodents as Reservoirs of Zoonotic Viruses: Are Bats Special? *Proceedings of the Royal Society B: Biological Sciences* 280(1756): 20122753.

Ma, Xiaoyue, Sarah Bonaparte, Patrick Corbett, Lillian A. Orciari, Crystal M. Gigante, Jordona D. Kirby, Richard B. Chipman, et al. 2023. Rabies Surveillance in the United States During 2021. *Journal of the American Veterinary Medical Association* 1:1–9.

Markotter, W., J. Coertse, L. De Vries, M. Geldenhuys, and M. Mortlock. 2020. Bat-Borne Viruses in Africa: A Critical Review. *Journal of Zoology* 311(2):77–98.

Mollentze, Nardus, and Daniel G. Streicker. 2020. Viral Zoonotic Risk Is Homogenous Among Taxonomic Orders of Mammalian and Avian Reservoir Hosts. *Proceedings of the National Academy of Sciences of the United States of America* 117(17):9423–9430.

Moreno, J. A., and G. M. Baer. 1980. Experimental Rabies in the Vampire Bat. *The American Journal of Tropical Medicine and Hygiene* 29(2):254–259.

Olival, Kevin J., Cristin C. Weekley, and Peter Daszak. 2015. Are Bats Really "Special" as Viral Reservoirs? What We Know and Need to Know. In *Bats and Viruses: A New Frontier of Emerging Infectious Diseases*. Edited by Lin-Fa Wang and Christopher Cowled. Hoboken, NJ: John Wiley and Sons. 281–294.

Olivero, Jesús, Julia E. Fa, Miguel Farfán, Ana L. Márquez, Raimundo Real, F. Javier Juste, Siv A. Leendertz, and Robert Nasi. 2020. Human Activities Link Fruit Bat Presence to Ebola Virus Disease Outbreaks. *Mammal Review* 50(1):1–10.

Rahman, Sohayati A., Sharifah S. Hassan, Kevin J. Olival, Maizan Mohamed, Li Yen Chang, Latiffah Hassan, Norsharina M. Saad, et al. 2010. Characterization of Nipah Virus from Naturally Infected *Pteropus vampyrus* Bats, Malaysia. *Emerging Infectious Diseases* 16(12):1990.

Ruiz-Aravena, Manuel, Clifton McKee, Amandine Gamble, Tamika Lunn, Aaron Morris, Celine E. Snedden, Claude Kwe Yinda, et al. 2021. Ecology, Evolution and Spillover of Coronaviruses from Bats. *Nature Reviews Microbiology* 2021 20:5 20(5):299–314.

Schatz, J., A. R. Fooks, L. McElhinney, D. Horton, J. Echevarria, S. Vázquez-Moron, E. A. Kooi, T. B. Rasmussen, T. Müller, and C. M. Freuling. 2013. Bat Rabies Surveillance in Europe. *Zoonoses and Public Health* 60(1):22–34.

Streicker, Daniel G., and Amy T. Gilbert. 2020. Contextualizing Bats as Viral Reservoirs. *Science* 370(6513):172–173.

Tan, Chee Wah, Xinglou Yang, Danielle E. Anderson, and Lin Fa Wang. 2021. Bat Virome Research: The Past, the Present and the Future. *Current Opinion in Virology* 49(August):68–80.

Visser, Marike, Rachelle Bester, Johan T. Burger, and Hans J. Maree. 2016. Next-Generation Sequencing for Virus Detection: Covering All the Bases. *Virology Journal* 13(1):1–6.

Woolhouse, Mark, Fiona Scott, Zoe Hudson, Richard Howey, and Margo Chase-Topping. 2012. Human Viruses: Discovery and Emergence. *Philosophical Transactions of the Royal Society B: Biological Sciences* 367(1604):2864.

# 12. Learn

Banerjee, Arinjay, Michelle L. Baker, Kirsten Kulcsar, Vikram Misra, Raina Plowright, and Karen Mossman. 2020. Novel Insights into Immune Systems of Bats. *Frontiers in Immunology* 11:507886.

Brunet-Rossinni, Anja K., and Steven N. Austad. 2004. Ageing Studies on Bats: A Review. *Biogerontology* 5(4):211–222.

Foley, Nicole M., Graham M. Hughes, Zixia Huang, Michael Clarke, David Jebb, Conor V. Whelan, Eric J. Petit, et al. 2018. Growing Old, Yet Staying Young: The Role of Telomeres in Bats' Exceptional Longevity. *Science Advances* 4(2):eaao0926.

Jebb, David, Zixia Huang, Martin Pippel, Graham M. Hughes, Ksenia Lavrichenko, Paolo Devanna, Sylke Winkler, et al. 2020. Six Reference-Quality Genomes Reveal Evolution of Bat Adaptations. *Nature* 583(7817):578–584.

Kelm, Detlev H., Ralph Simon, Doreen Kuhlow, Christian C. Voigt, and Michael Ristow. 2011. High Activity Enables Life on a High-Sugar Diet: Blood Glucose Regulation in Nectar-Feeding Bats. *Proceedings of the Royal Society B: Biological Sciences* 278(1724):3490–3496.

Knörnschild, Mirjam. 2014. Vocal Production Learning in Bats. *Current Opinion in Neurobiology* 28:80–85.

Lagunas-Rangel, Francisco Alejandro. 2020. Why Do Bats Live So Long? Possible Molecular Mechanisms. *Biogerontology* 21(1):1–11.

Lilley, T. M., J. Stauffer, M. Kanerva, and T. Eeva. 2014. Interspecific Variation in Redox Status Regulation and Immune Defence in Five Bat Species: The Role of Ectoparasites. *Oecologia* 175(3):811–823.

Mansour, Carl Bou, Elijah Koreman, Dennis Laurijssen, Jan Steckel, Herbert Peremans, and Dieter Vanderelst. 2019. Robotic Models of Obstacle Avoidance in Bats. *Artificial Life Conference Proceedings*, 463–464.

Moreno Santillán, Diana D., Tanya M. Lama, Yocelyn T. Gutierrez Guerrero, Alexis M. Brown, Paul Donat, Huabin Zhao, Stephen J. Rossiter, et al. 2021. Large-Scale Genome Sampling Reveals Unique Immunity and Metabolic Adaptations in Bats. *Molecular Ecology* 30(23):6449–6467.

Potter, Joshua H. T., Rosie Drinkwater, Kalina T. J. Davies, Nicolas Nesi, Marisa C. W. Lim, Laurel R. Yohe, Hai Chi, et al. 2021. Nectar-Feeding Bats and Birds Show Parallel Molecular Adaptations in Sugar Metabolism Enzymes. *Current Biology* 31(20):4667–4674.e6.

Power, Megan L., Nicole M. Foley, Gareth Jones, and Emma C. Teeling. 2022. Taking Flight: An Ecological, Evolutionary and Genomic Perspective on Bat Telomeres. *Molecular Ecology* 31(23):6053–6068.

Ramezani, Alireza, Soon Jo Chung, and Seth Hutchinson. 2017. A Biomimetic Robotic Platform to Study Flight Specializations of Bats. *Science Robotics* 2(3):eaal2505.

Salmon, Adam B., Shanique Leonard, Venkata Masamsetti, Anson Pierce, Andrej J. Podlutsky, Natalia Podlutskaya, Arlan Richardson, Steven N. Austad, and Asish R. Chaudhuri. 2009. The Long Lifespan of Two Bat Species Is Correlated with Resistance to Protein Oxidation and Enhanced Protein Homeostasis. *The FASEB Journal* 23(7):2317.

Schneeberger, Karin, Gábor Á Czirják, and Christian C. Voigt. 2014. Frugivory Is Associated with Low Measures of Plasma Oxidative Stress and High Antioxidant Concentration in Free-Ranging Bats. *Naturwissenschaften* 101(4):285–290.

Seim, Inge, Xiaodong Fang, Zhiqiang Xiong, Alexey V. Lobanov, Zhiyong Huang, Siming Ma, Yue Feng, et al. 2013. Genome Analysis Reveals Insights into Physiology and Longevity of the Brandt's Bat *Myotis brandtii*. *Nature Communications* 4(1):1–8.

Shen, Yong Yi, Lu Liang, Zhou Hai Zhu, Wei Ping Zhou, David M. Irwin, and Ya Ping Zhang. 2010. Adaptive Evolution of Energy Metabolism Genes and the Origin of Flight in Bats. *Proceedings of the National Academy of Sciences of the United States of America* 107(19):8666–8671.

Sohl-Dickstein, Jascha, Benjamin M. Gaub, Chris C. Rodgers, Crystal Li, Michael R Deweese, and Nicol S. Harper. 2015. A Device for Human Ultrasonic Echolocation. *IEEE Transactions in Biomedical Engineering* 62(6):1526–1534.

Sullivan, Isabel R., Danielle M. Adams, Lucas J. S. Greville, Paul A. Faure, and Gerald S. Wilkinson. 2022. Big Brown Bats Experience Slower Epigenetic Ageing During Hibernation. *Proceedings of the Royal Society B: Biological Sciences* 289(1980):20220635.

Teeling, Emma C., Sonja C. Vernes, Liliana M. Dávalos, David A. Ray, M. Thomas, P. Gilbert, Eugene Myers, and Bat1K Consortium. 2018. Bat Biology, Genomes, and the Bat1K Project: To Generate Chromosome-Level Genomes for All Living Bat Species. *Annual Review of Animal Biosciences* 6:23–46.

Thaler, Lore, and Melvyn A. Goodale. 2016. Echolocation in Humans: An Overview. *Cognitive Science* 7(6):382–393.

Tian, Shilin, Jiaming Zeng, Hengwu Jiao, Dejing Zhang, Libiao Zhang, Cao-Qi Lei, Stephen J. Rossiter, and Huabin Zhao. 2023. Comparative Analyses of Bat Genomes Identify Distinct Evolution of Immunity in Old World Fruit Bats. *Science Advances* 9:eaddo141.

Vernes, Sonja C. 2017. What Bats Have to Say About Speech and Language. *Psychonomic Bulletin and Review* 24(1):111–117.

Vernes, Sonja C., and Gerald S. Wilkinson. 2020. Behaviour, Biology and Evolution of Vocal Learning in Bats. *Philosophical Transactions of the Royal Society B* 375(1789):20190061.

Vernes, Sonja C., Paolo Devanna, Stephen Gareth Hörpel, Ine Alvarez van Tussenbroek, Uwe Firzlaff, Peter Hagoort, Michael Hiller, et al. 2022. The Pale Spear-Nosed Bat: A Neuromolecular and Transgenic Model for Vocal Learning. *Annals of the New York Academy of Sciences* 1517(1):125–142.

Voigt, C. C., and J. R. Speakman. 2007. Nectar-Feeding Bats Fuel Their High Metabolism Directly with Exogenous Carbohydrates. *Functional Ecology* 21(5):913–921.

Wilkinson, Gerald S., and Danielle M. Adams. 2019. Recurrent Evolution of Extreme Longevity in Bats. *Biology Letters* 15(4):20180860.

Wu, Cheng Wei, and Kenneth B. Storey. 2016. Life in the Cold: Links Between Mammalian Hibernation and Longevity. *Biomolecular Concepts* 7(1):41–52.

Zhang, Guojie, Christopher Cowled, Zhengli Shi, Zhiyong Huang, Kimberly A. Bishop-Lilly, Xiaodong Fang, James W. Wynne, et al. 2013. Comparative Analysis of Bat Genomes Provides Insight into the Evolution of Flight and Immunity. *Science* 339(6118):456–460.

# 13. Keystones

Abedi-Lartey, Michael, Dina K. N. Dechmann, Martin Wikelski, Anne K. Scharf, and Jakob Fahr. 2016. Long-Distance Seed Dispersal by Straw-Coloured Fruit Bats Varies by Season and Landscape. *Global Ecology and Conservation* 7:12–24.

Aguiar, Ludmilla M. S., Igor D. Bueno-Rocha, Guilherme Oliveira, Eder S. Pires, Santelmo Vasconcelos, Gisele L. Nunes, Marina R. Frizzas, and Pedro H. B. Togni. 2021. Going Out for Dinner: The Consumption of Agriculture Pests by Bats in Urban Areas. *PLoS ONE* 16(10):e0258066.

Alpízar, Priscilla, Julian Schneider, and Marco Tschapka. 2020. Bats and Bananas: Simplified Diet of the Nectar-Feeding Bat *Glossophaga soricina* (Phyllostomidae: Glossophaginae) Foraging in Costa Rican Banana Plantations. *Global Ecology and Conservation* 24(December):e01254.

Ancillotto, Leonardo, Rosario Rummo, Giulia Agostinetto, Nicola Tommasi, Antonio P. Garonna, Flavia de Benedetta, Umberto Bernardo, Andrea Galimberti, and Danilo Russo. 2022. Bats as Suppressors of Agroforestry Pests in Beech Forests. *Forest Ecology and Management* 522(October):120467.

Arias-Cóyotl, Ethel, Kathryn E. Stoner, and Alejandro Casas. 2006. Effectiveness of Bats as Pollinators of *Stenocereus stellatus* (Cactaceae) in Wild, Managed in Situ, and Cultivated Populations in La Mixteca Baja, Central Mexico. *American Journal of Botany* 93(11):1675–1683.

Aziz, Sheema A., Gopalasamy R. Clements, Kim R. McConkey, Tuanjit Sritongchuay, Saifful Pathil, Muhammad Nur Hafizi Abu Yazid, Ahimsa Campos-Arceiz, Pierre Michel Forget, and Sara Bumrungsri. 2017. Pollination by the Locally Endangered Island Flying Fox (*Pteropus hypomelanus*) Enhances Fruit Production of the Economically Important Durian (*Durio zibethinus*). *Ecology and Evolution* 7(21):8670–8684.

Baqi, Aminuddin, Voon-Ching Lim, Hafiz Yazid, Faisal Ali Anwarali Khan, Chong Ju Lian, Bryan Raveen Nelson, Jaya Seelan Sathiya Seelan, Suganthi Appalasamy, Seri Intan Mokhtar, and Jayaraj Vijaya Kumaran. 2022. A Review of Durian Plant–Bat Pollinator Interactions. *Journal of Plant Interactions* 17(1):105–126.

Beilke, Elizabeth A., and Joy M. O'Keefe. 2023. Bats Reduce Insect Density and Defoliation in Temperate Forests: An Exclusion Experiment. *Ecology* 104(2):e3903.

Bhalla, Iqbal Singh, Jesús Aguirre Gutiérrez, and Robert J. Whittaker. 2023. Batting for Rice: The Effect of Bat Exclusion on Rice in North-East India. *Agriculture, Ecosystems & Environment* 341:108196.

Boyles, Justin, and Catherine L Sole. 2013. On Estimating the Economic Value of Insectivorous Bats: Prospects and Priorities for Biologists. In *Bat Evolution, Ecology, and Conservation*. Edited by Rick A. Adams and Scott C. Pedersen. New York: Springer. 501–515.

Classen, Alice, Marcell K. Peters, Stefan W. Ferger, Maria Helbig-Bonitz, Julia M. Schmack, Genevieve Maassen, Matthias Schleuning, Elisabeth K. V. Kalko, Katrin Böhning-Gaese, and Ingolf Steffan-Dewenter. 2014. Complementary Ecosystem Services Provided by Pest Predators and Pollinators Increase Quantity and Quality of Coffee Yields. *Proceedings of the Royal Society B: Biological Sciences* 281(1779):20133148.

Djossa, Bruno Agossou, Hermann Cyr Toni, Ibrahim Dende Adekanmbi, Florida K Tognon, and Brice Augustin Sinsin. 2015. Do Flying Foxes Limit Flower Abortion in African Baobab (*Adansonia digitata*)? Case Study in Benin, West Africa. *EDP Sciences* 70(5):281–287.

Federico, Paula, Thomas G. Hallam, Gary F. McCracken, S. Thomas Purucker, William E. Grant, A. Nelly Correa-Sandoval, John K. Westbrook, et al. 2008. Brazilian Free-Tailed Bats as Insect Pest Regulators in Transgenic and Conventional Cotton Crops. *Ecological Applications* 18(4):826–837.

Fenolio, Danté B., G. O. Graening, Bret A. Collier, and Jim F. Stout. 2005. Coprophagy in a Cave-Adapted Salamander: The Importance of Bat Guano Examined Through Nutritional and Stable Isotope Analyses. *Proceedings of the Royal Society B: Biological Sciences* 273(1585):439–443.

Frick, Winifred F., Paul A. Heady, and John P. Hayes. 2009. Facultative Nectar-Feeding Behavior in a Gleaning Insectivorous Bat (*Antrozous pallidus*). *Journal of Mammalogy* 90(5):1157–1164.

Frick, Winifred F., Ryan D. Price, Paul A. Heady, and Kathleen M. Kay. 2013. Insectivorous Bat Pollinates Columnar Cactus More Effectively per Visit than Specialized Nectar Bat. *The American Naturalist* 181(1):137–144.

Griffin, Donald R., Frederic A. Webster, and Charles R. Michael. 1960. The Echolocation of Flying Insects by Bats. *Animal Behaviour* 8(3–4):141–154.

Guero, Yadji, Ambouta Harouna Karimou, Guero Yadji, Abdou Gado Fanna, and Abarchi Idrissa. 2020. Effect of Different Rate of Bat Guano on Growth and Yield of Tomatoes (*Lycopersicon esculentum* Mill.) in Niamey, Niger. *Journal of Experimental Agriculture International* 42(3):34–46.

Hodgkison, Robert, Sharon T. Balding, Akbar Zubaid, and Thomas H. Kunz. 2003. Fruit Bats (Chiroptera: Pteropodidae) as Seed Dispersers and Pollinators in a Lowland Malaysian Rain Forest. *Biotropica* 35(4):491–502.

Hughes, Morgan J., Elizabeth C. Braun de Torrez, Eva A. Buckner, and Holly K. Ober. 2022. Consumption of Endemic Arbovirus Mosquito Vectors by Bats in the Southeastern United States. *Journal of Vector Ecology* 47(2):153–165.

Itino, Takao, Makoto Kato, and Mitsuru Hotta. 1991. Pollination Ecology of the Two Wild Bananas, *Musa acuminata halabanensis* and *M. salaccensis*: Chiropterophily and Ornithophily. *Biotropica* 23(2):151.

Kalka, Margareta B., Adam R. Smith, and Elisabeth K. V. Kalko. 2008. Bats Limit Arthropods and Herbivory in a Tropical Forest. *Science* 320(5872):71.

Kolkert, Heidi, Rhiannon Smith, Romina Rader, and Nick Reid. 2021. Insectivorous Bats Provide Significant Economic Value to the Australian Cotton Industry. *Ecosystem Services* 49(June):101280.

Kunz, Thomas H., Elizabeth Braun de Torrez, Dana Bauer, Tatyana Lobova, and Theodore H. Fleming. 2011. Ecosystem Services Provided by Bats. *Annals of the New York Academy of Sciences* 1223(1):1–38.

Law, Bradley S., and Merrilyn Lean. 1999. Common Blossom Bats (*Syconycteris australis*) as Pollinators in Fragmented Australian Tropical Rainforest. *Biological Conservation* 91(2–3):201–212.

Leelapaibul, Watcharee, Sara Bumrungsri, and Anak Pattanawiboon. 2005. Diet of Wrinkle-Lipped Free-Tailed Bat (*Tadarida plicata* Buchannan, 1800) in Central Thailand: Insectivorous Bats Potentially Act as Biological Pest Control Agents. *Acta Chiropterologica* 7(1):111–119.

Maas, Bea, Daniel S. Karp, Sara Bumrungsri, Kevin Darras, David Gonthier, Joe C. C. Huang, Catherine A. Lindell, et al. 2016. Bird and Bat Predation Services in Tropical Forests and Agroforestry Landscapes. *Biological Reviews* 91(4):1081–1101.

Mainea, Josiah J., and Justin G. Boyles. 2015. Bats Initiate Vital Agroecological Interactions in Corn. *Proceedings of the National Academy of Sciences of the United States of America* 112(40):12438–12443.

Maslo, Brooke, Rebecca L. Mau, Kathleen Kerwin, Ryelan McDonough, Erin McHale, and Jeffrey T. Foster. 2022. Bats Provide a Critical Ecosystem Service by Consuming a Large Diversity of Agricultural Pest Insects. *Agriculture, Ecosystems & Environment* 324(February):107722.

Melo, Felipe P. L., Bernal Rodriguez-Herrera, Robin L. Chazdon, Rodrigo A. Medellin, and Gerardo G. Ceballos. 2009. Small Tent-Roosting Bats Promote Dispersal of Large-Seeded Plants in a Neotropical Forest. *Biotropica* 41(6):737–743.

Morrison, Emily B., and Catherine A. Lindell. 2012. Birds and Bats Reduce Insect Biomass and Leaf Damage in Tropical Forest Restoration Sites. *Ecological Applications* 22(5):1526–1534.

Murphy, Megan, Elizabeth L. Clare, Jens Rydell, Yossi Yovel, Yinon Bar-On, Phillip J. Oelbaum, and M. Brock Fenton. 2016. Opportunistic Use of Banana Flower Bracts by *Glossophaga soricina*. *Acta Chiropterologica* 18(1):209–213.

Newman, Ethan, Keeveshnee Govender, Sandy van Niekerk, and Steven D. Johnson. 2021. The Functional Ecology of Bat Pollination in the African Sausage Tree *Kigelia africana* (Bignoniaceae). *Biotropica* 53(2):477–486.

Nor Zalipah, Mohamed, Mohd Sah Shahrul Anuar, and Gareth Jones. 2016. The Potential Significance of Nectar-Feeding Bats as Pollinators in Mangrove Habitats of Peninsular Malaysia. *Biotropica* 48(4):425–428.

Puig-Montserrat, Xavier, Ignasi Torre, Adrià López-Baucells, Emilio Guerrieri, Maurilia M. Monti, Ruth Ràfols-García, Xavier Ferrer, David Gisbert, and Carles Flaquer. 2015. Pest Control Service Provided by Bats in Mediterranean Rice Paddies: Linking Agroecosystems Structure to Ecological Functions. *Mammalian Biology* 80:237–245.

Reiskind, Michael H., and Matthew A. Wund. 2009. Experimental Assessment of the Impacts of Northern Long-Eared Bats on Ovipositing *Culex* (Diptera: Culicidae) Mosquitoes. *Journal of Medical Entomology* 46(5):1037–1044.

Rodríguez-San Pedro, Annia, Juan Luis Allendes, Clemente A. Beltrán, Pascal N. Chaperon, Mónica M. Saldarriaga-Córdoba, Andrea X. Silva, and Audrey A. Grez. 2020. Quantifying Ecological and Economic Value of Pest Control Services Provided by Bats in a Vineyard Landscape of Central Chile. *Agriculture, Ecosystems & Environment* 302(October):107063.

Rydell, Jens, Doreen Parker McNeill, and Johan Eklöf. 2002. Capture Success of Little Brown Bats (*Myotis lucifugus*) Feeding on Mosquitoes. *Journal of Zoology* 256(3):379–381.

Sakoui, Souraya, Reda Derdak, Boutaina Addoum, Aurelio Serrano-Delgado, Abdelaziz Soukri, and Bouchra El Khalfi. 2020. The Life Hidden Inside Caves: Ecological and Economic Importance of Bat Guano. *International Journal of Ecology* 2020:9872532.

Salazar, Diego, Detlev H. Kelm, and Robert J. Marquis. 2013. Directed Seed Dispersal of *Piper* by *Carollia perspicillata* and Its Effect on Understory Plant Diversity and Folivory. *Ecology* 94(11):2444–2453.

Seltzer, Carrie E., Henry J. Ndangalasi, and Norbert J. Cordeiro. 2013. Seed Dispersal in the Dark: Shedding Light on the Role of Fruit Bats in Africa. *Biotropica* 45(4):450–456.

Sheherazade, Holly K. Ober, and Susan M. Tsang. 2019. Contributions of Bats to the Local Economy Through Durian Pollination in Sulawesi, Indonesia. *Biotropica* 51(6):913–922.

Srilopan, Supawan, Sara Bumrungsri, and Sopark Jantarit. 2018. The Wrinkle-Lipped Free-Tailed Bat (*Chaerephon plicatus* Buchannan, 1800) Feeds Mainly on Brown Planthoppers in Rice Fields of Central Thailand. *Acta Chiropterologica* 20(1):207–219.

Taylor, Peter John, Ingo Grass, Andries J. Alberts, Elsje Joubert, and Teja Tscharntke. 2018. Economic Value of Bat Predation Services: A Review and New Estimates from Macadamia Orchards. *Ecosystem Services* 30:372–381.

Taylor, Peter J., Catherine Vise, Macy A. Krishnamoorthy, Tigga Kingston, and Sarah Venter. 2020. Citizen Science Confirms the Rarity of Fruit Bat Pollination of Baobab (*Adansonia digitata*) Flowers in Southern Africa. *Diversity* 12(3):106.

Tremlett, Constance J., Mandy Moore, Mark A. Chapman, Veronica Zamora-Gutierrez, and Kelvin S.-H Peh. 2020. Pollination by Bats Enhances Both Quality and Yield of a Major Cash Crop in Mexico. *Journal of Applied Ecology* 57:450–459.

Tremlett, Constance J., Kelvin S. H. Peh, Veronica Zamora-Gutierrez, and Marije Schaafsma. 2021. Value and Benefit Distribution of Pollination Services Provided by Bats in the Production of Cactus Fruits in Central Mexico. *Ecosystem Services* 47:101197.

Wanger, Thomas Cherico, Kevin Darras, Sara Bumrungsri, Teja Tscharntke, and Alexandra Maria Klein. 2014. Bat Pest Control Contributes to Food Security in Thailand. *Biological Conservation* 171(March):220–223.

Wetzler, Gabrielle C., and Justin G. Boyles. 2018. The Energetics of Mosquito Feeding by Insectivorous Bats. *Canadian Journal of Zoology* 96(4):373–377.

Williams-Guillén, Kimberly, Ivette Perfecto, and John Vandermeer. 2008. Bats Limit Insects in a Neotropical Agroforestry System. *Science* 320(5872):70.

Wray, Amy K., Michelle A. Jusino, Mark T. Banik, Jonathan M. Palmer, Heather Kaarakka, J. Paul White, Daniel L. Lindner, Claudio Gratton, and M. Zachariah Peery. 2018. Incidence and Taxonomic Richness of Mosquitoes in the Diets of Little Brown and Big Brown Bats. *Journal of Mammalogy* 99(3):668–674.

Zalipah, Mohamed Nor, Mohd Sah Shahrul Anuar, and Gareth Jones. 2016. The Potential Significance of Nectar-Feeding Bats as Pollinators in Mangrove Habitats of Peninsular Malaysia. *Biotropica* 48(4):425 428.

# 14. Conserve

Alducin-Martínez, Cecilia, Karen Y. Ruiz Mondragón, Ofelia Jiménez-Barrón, Erika Aguirre-Planter, Jaime Gasca-Pineda, Luis E. Eguiarte, and Rodrigo A. Medellin. 2023. Uses, Knowledge and Extinction Risk Faced by *Agave* Species in Mexico. *Plants* 12(1):124.

Allen, Louise C., Jesse R. Barber, Nickolay I. Hristov, Juliette J. Rubin, and Joseph T. Lightsey. 2021. Noise Distracts Foraging Bats. *Proceedings of Royal Society B: Biological Sciences* 288(1944):20202689.

Allinson, Graeme, Cindi Mispagel, Natsuko Kajiwara, Yasumi Anan, Junko Hashimoto, Laurie Laurenson, Mayumi Allinson, and Shinsuke Tanabe. 2006. Organochlorine and Trace Metal Residues in Adult Southern Bent-Wing Bat (*Miniopterus schreibersii bassanii*) in Southeastern Australia. *Chemosphere* 64(9):1464–1471.

Ancillotto, Leonardo, Maria Tiziana Serangeli, and Danilo Russo. 2013. Curiosity Killed the Bat: Domestic Cats as Bat Predators. *Mammalian Biology* 78(5):369–373.

Anderson, A., S. Shwiff, K. Gebhardt, A. J. Ramírez, S. Shwiff, D. Kohler, and L. Lecuona. 2014. Economic Evaluation of Vampire Bat (*Desmodus rotundus*) Rabies Prevention in Mexico. *Transboundary and Emerging Diseases* 61(2):140–146.

Arnett, Edward B., Erin F. Baerwald, Fiona Mathews, Luisa Rodrigues, Armando Rodríguez-Durán, Jens Rydell, Rafael Villegas-Patraca, and Christian C. Voigt. 2015. Impacts of Wind Energy Development on Bats: A Global Perspective. In *Bats in the Anthropocene: Conservation of Bats in a Changing World*. Edited by Christian C. Voigt and Tigga Kingston. New York: Springer. 295–323.

Auteri, Giorgia G., and L. Lacey Knowles. 2020. Decimated Little Brown Bats Show Potential for Adaptive Change. *Scientific Reports* 10(1):1–10.

Baerwald, Erin F., Genevieve H. D'Amours, Brandon J. Klug, and Robert M. R. Barclay. 2008. Barotrauma Is a Significant Cause of Bat Fatalities at Wind Turbines. *Current Biology* 18(16):R695–696.

Barber, Jesse R., Kevin R. Crooks, and Kurt M. Fristrup. 2010. The Costs of Chronic Noise Exposure for Terrestrial Organisms. *Trends in Ecology & Evolution* 25(3):180–189.

Bayat, Sara, Fritz Geiser, Paul Kristiansen, and Susan C. Wilson. 2014. Organic Contaminants in Bats: Trends and New Issues. *Environment International* 63:40–52.

Blehert, David S., Alan C. Hicks, Melissa Behr, Carol U. Meteyer, Brenda M. Berlowski-Zier, Elizabeth L. Buckles, Jeremy T. H. Coleman, et al. 2009. Bat White-Nose Syndrome: An Emerging Fungal Pathogen? *Science* 323(5911):227.

Boso, Àlex, Boris Álvarez, Beatriz Pérez, Juan Carlos Imio, Adison Altamirano, and Fulgencio Lisón. 2021. Understanding Human Attitudes Towards Bats and the Role of Information and Aesthetics to Boost a Positive Response as a Conservation Tool. *Animal Conservation* 24(6):937–945.

Buchweitz, John P., Keri Carson, Sarah Rebolloso, and Andreas Lehner. 2018. DDT Poisoning of Big Brown Bats, *Eptesicus fuscus*, in Hamilton, Montana. *Chemosphere* 201(June):1–5.

Bunkley, Jessie Patrice, and Jesse Rex Barber. 2015. Noise Reduces Foraging Efficiency in Pallid Bats (*Antrozous pallidus*). *Ethology* 121(11):1116–1121.

Cable, Ashleigh B., Emma V. Willcox, and Christy Leppanen. 2021. Contaminant Exposure as an Additional Stressor to Bats Affected by White-Nose Syndrome: Current Evidence and Knowledge Gaps. *Ecotoxicology* 31(1):12–23.

Campana, Michael G., Naoko P. Kurata, Jeffrey T. Foster, Lauren E. Helgen, Dee Ann M. Reeder, Robert C. Fleischer, and Kristofer M. Helgen. 2017. White-Nose Syndrome Fungus in a 1918 Bat Specimen from France. *Emerging Infectious Diseases* 23(9):1611–1612.

Chaber, Anne Lise, Kyle N. Amstrong, Sigit Wiantoro, Vanessa Xerri, Charles Caraguel, Wayne S. J. Boardman, and Torben D. Nielsen. 2021. Bat E-Commerce: Insights into the Extent and Potential Implications of This Dark Trade. *Frontiers in Veterinary Science* 8(June):543.

Cheng, Tina L., Heather Mayberry, Liam P. McGuire, Joseph R. Hoyt, Kate E. Langwig, Hung Nguyen, Katy L. Parise, et al. 2017. Efficacy of a Probiotic Bacterium to Treat Bats Affected by the Disease White-Nose Syndrome. *Journal of Applied Ecology* 54(3):701–708.

Cheng, Tina L., Alexander Gerson, Marianne S. Moore, Jonathan D. Reichard, Joely DeSimone, Craig K. R. Willis, Winifred F. Frick, and Auston Marm Kilpatrick. 2019. Higher Fat Stores Contribute to Persistence of Little Brown Bat Populations with White-Nose Syndrome. *Journal of Animal Ecology* 88(4):591–600.

Cheng, Tina L., Jonathan D. Reichard, Jeremy T. H. Coleman, Theodore J. Weller, Wayne E. Thogmartin, Brian E. Reichert, Alyssa B. Bennett, et al. 2021. The Scope and Severity of White-Nose Syndrome on Hibernating Bats in North America. *Conservation Biology* 35(5):1586–1597.

Clark, D. R. 2001. DDT and the Decline of Free-Tailed Bats (*Tadarida brasiliensis*) at Carlsbad Cavern, New Mexico. *Archives of Environmental Contamination and Toxicology* 40(4):537–543.

Crawford, Reed D., Luke E. Dodd, Francis E. Tillman, and Joy M. O'Keefe. 2022. Evaluating Bat Boxes: Design and Placement Alter Bioenergetic Costs and Overheating Risk. *Conservation Physiology* 10(1):coac027.

Diengdoh, Vishesh L., Stefania Ondei, Mark Hunt, and Barry W. Brook. 2022. Predicted Impacts of Climate Change and Extreme Temperature Events on the Future Distribution of Fruit Bat Species in Australia. *Global Ecology and Conservation* 37(September):e02181.

Festa, Francesca, Leonardo Ancillotto, Luca Santini, Michela Pacifici, Ricardo Rocha, Nia Toshkova, Francisco Amorim, et al. 2023. Bat Responses to Climate Change: A Systematic Review. *Biological Reviews* 98(1):19–33.

Flaquer, Carles, Ignacio Torre, and Ramon Ruiz-Jarillo. 2006. The Value of Bat-Boxes in the Conservation of *Pipistrellus pygmaeus* in Wetland Rice Paddies. *Biological Conservation* 128(2):223–230.

Frick, Winifred F., Jacob F. Pollock, Alan C. Hicks, Kate E. Langwig, D. Scott Reynolds, Gregory G. Turner, Calvin M. Butchkoski, and Thomas H. Kunz. 2010. An Emerging Disease Causes Regional Population Collapse of a Common North American Bat Species. *Science* 329(5992):679–682.

Frick, Winifred F., E. F. Baerwald, J. F. Pollock, R. M. R. Barclay, J. A. Szymanski, T. J. Weller, A. L. Russell, S. C. Loeb, R. A. Medellin, and L. P. McGuire. 2017. Fatalities at Wind Turbines May Threaten Population Viability of a Migratory Bat. *Biological Conservation* 209(May):172–177.

Frick, Winifred F., Tigga Kingston, and Jon Flanders. 2020. A Review of the Major Threats and Challenges to Global Bat Conservation. *Annals of the New York Academy of Sciences* 1469(1):5–25.

Frick, Winifred F., Emily Johnson, Tina L. Cheng, Julia S. Lankton, Robin Warne, Jason Dallas, Katy L. Parise, Jeffrey T. Foster, Justin G. Boyles, and Liam P. McGuire. 2022. Experimental Inoculation Trial to Determine the Effects of Temperature and Humidity on White-Nose Syndrome in Hibernating Bats. *Scientific Reports* 12(1):1–13.

Frick, Winifred F., Yvonne A. Dzal, Kristin A. Jonasson, Michael D. Whitby, Amanda M. Adams, Christen Long, John E. Depue, et al. 2023. Bats Increased Foraging Activity at Experimental Prey Patches near Hibernacula. *Ecological Solutions and Evidence* 4(11):e12217.

Friedenberg, Nicholas A., and Winifred F. Frick. 2021. Assessing Fatality Minimization for Hoary Bats amid Continued Wind Energy Development. *Biological Conservation* 262(October):109309.

Fuller, Nathan W., Liam P. McGuire, Evan L. Pannkuk, Todd Blute, Catherine G. Haase, Heather W. Mayberry, Thomas S. Risch, and Craig K. R. Willis. 2020. Disease Recovery in Bats Affected by White-Nose Syndrome. *Journal of Experimental Biology* 223(6):jeb211912.

Furey, Neil M., and Paul A. Racey. 2015. Conservation Ecology of Cave Bats. In *Bats in the Anthropocene: Conservation of Bats in a Changing World*. Edited by Christian C. Voigt and Tigga Kingston. New York: Springer. 463–500.

Gabriel, Kyle T., Ashley G. McDonald, Kelly E. Lutsch, Peter E. Pattavina, Katrina M. Morris, Emily A. Ferrall, Sidney A. Crow, and Christopher T. Cornelison. 2022. Development of a Multi-Year White-Nose Syndrome Mitigation Strategy Using Antifungal Volatile Organic Compounds. *PLoS ONE* 17(12):e0278603.

Gignoux-Wolfsohn, Sarah A., Malin L. Pinsky, Kathleen Kerwin, Carl Herzog, MacKenzie Hall, Alyssa B. Bennett, Nina H. Fefferman, and Brooke Maslo. 2021. Genomic Signatures of Selection in Bats Surviving White-Nose Syndrome. *Molecular Ecology* 30(22):5643–5657.

Gómez-Ruiz, Emma P., and Thomas E. Lacher. 2019. Climate Change, Range Shifts, and the Disruption of a Pollinator–Plant Complex. *Scientific Reports* 9(1):1–10.

Griffiths, Stephen R., Robert Bender, Lisa N. Godinho, Pia E. Lentini, Linda F. Lumsden, and Kylie A. Robert. 2017. Bat Boxes Are Not a Silver Bullet Conservation Tool. *Mammal Review* 47(4):261–265.

Grodsky, Steven M., Melissa J. Behr, Andrew Gendler, David Drake, Byron D. Dieterle, Robert J. Rudd, and Nicole L. Walrath. 2011. Investigating the Causes of Death for Wind Turbine–Associated Bat Fatalities. *Journal of Mammalogy* 92(5):917–925.

Hayes, Mark A. 2013. Bats Killed in Large Numbers at United States Wind Energy Facilities. *BioScience* 63(12):975–979.

Hayes, Mark A., Lauren A. Hooton, Karen L. Gilland, Chuck Grandgent, Robin L. Smith, Stephen R. Lindsay, Jason D. Collins, et al. 2019. A Smart Curtailment

Approach for Reducing Bat Fatalities and Curtailment Time at Wind Energy Facilities. *Ecological Applications* 29(4):e01881.

Heim, Olga, Julia T. Treitler, Marco Tschapka, Mirjam Knörnschild, and Kirsten Jung. 2015. The Importance of Landscape Elements for Bat Activity and Species Richness in Agricultural Areas. *PLoS ONE* 10(7):e0134443.

Hoyt, Joseph R., Tina L. Cheng, Kate E. Langwig, Mallory M. Hee, Winifred F. Frick, and A. Marm Kilpatrick. 2015. Bacteria Isolated from Bats Inhibit the Growth of *Pseudogymnoascus destructans*, the Causative Agent of White-Nose Syndrome. *PLoS ONE* 10(4):e0121329.

Hoyt, Joseph R., A. Marm Kilpatrick, and Kate Langwig. 2021. Ecology and Impacts of White-Nose Syndrome on Bats. *Nature Reviews Microbiology* 19(3):196–210.

Hsiao, Chun Jen, Ching Lung Lin, Tian Yu Lin, Sheue Er Wang, and Chung Hsin Wu. 2016. Imidacloprid Toxicity Impairs Spatial Memory of Echolocation Bats Through Neural Apoptosis in Hippocampal CA1 and Medial Entorhinal Cortex Areas. *NeuroReport* 27(6):462–468.

Johnson, Joseph S., John J. Treanor, Alexandra C. Slusher, and Michael J. Lacki. 2019. Buildings Provide Vital Habitat for Little Brown Myotis (*Myotis lucifugus*) in a High-Elevation Landscape. *Ecosphere* 10(11):e02925.

Johnson, Laura, and Eluned C. Price. 2023. Battitude: A Virtual Zoo "Bat Experience" Produces Positive Change in Attitudes to an Unpopular Species. *Journal of Zoo and Aquarium Research* 11(1):232–239.

Jung, Kirsten, and Caragh G. Threlfall. 2015. Urbanisation and Its Effects on Bats: A Global Meta-Analysis. In *Bats in the Anthropocene: Conservation of Bats in a Changing World*. Edited by Christian C. Voigt and Tigga Kingston. New York: Springer. 13–33.

Kannan, Kurunthachalam, Se Hun Yun, Robert J. Rudd, and Melissa Behr. 2010. High Concentrations of Persistent Organic Pollutants Including PCBs, DDT, PBDEs and PFOS in Little Brown Bats with White-Nose Syndrome in New York, USA. *Chemosphere* 80(6):613–618.

Kingston, Tigga. 2015. Cute, Creepy, or Crispy: How Values, Attitudes, and Norms Shape Human Behavior Toward Bats. In *Bats in the Anthropocene: Conservation of Bats in a Changing World*. Edited by Christian C. Voigt and Tigga Kingston. New York: Springer. 571–595.

Langwig, Kate E., Joseph R. Hoyt, Katy L. Parise, Winifred F. Frick, Jeffrey T. Foster, and A. Marm Kilpatrick. 2017. Resistance in Persisting Bat Populations After White-Nose Syndrome Invasion. *Philosophical Transactions of the Royal Society B: Biological Sciences* 372(1712):20160044.

Lausen, Cori L., Pia Lentini, Susan Dulc, Leah Rensel, Caragh G. Threlfall, Emily de Freitas, and Mandy Kellner. 2022. Bat Boxes as Roosting Habitat in Urban Centres: "Thinking Outside the Box." In *Urban Bats: Biology, Ecology, and Human Dimensions*. Edited by Lauren Moretto, Joanna L. Coleman, Christina M. Davy, M. Brock Fenton, Carmi Korine, and Krista J. Patriquin. New York: Springer. 75–93.

Law, Bradley, Kirsty J. Park, and Michael J. Lacki. 2015. Insectivorous Bats and Silviculture: Balancing Timber Production and Bat Conservation. In *Bats in the Anthropocene: Conservation of Bats in a Changing World*. Edited by Christian C. Voigt and Tigga Kingston. New York: Springer. 105–150.

Lee, Benjamin P.Y.-H., Matthew J. Struebig, Stephen J Rossiter, and Tigga Kingston. 2015. Increasing Concern over Trade in Bat Souvenirs from South-East Asia. *Oryx* 49(2):204.

Leopardi, Stefania, Damer Blake, and Sébastien J. Puechmaille. 2015. White-Nose Syndrome Fungus Introduced from Europe to North America. *Current Biology* 25(6):R217–219.

Lintott, Paul R., and Fiona Mathews. 2018. *Reviewing the Evidence on Mitigation Strategies for Bats in Buildings: Informing Best-Practice for Policy Makers and Practitioners*. Hampshire, UK: Chartered Institute of Ecology and Environmental Management.

López-Baucells, Adrià, Ricardo Rocha, and Álvaro Fernández-Llamazares. 2018. When Bats Go Viral: Negative Framings in Virological Research Imperil Bat Conservation. *Mammal Review* 48(1):62–66.

McCracken, Gary F., Riley F. Bernard, Melquisidec Gamba-Rios, Randy Wolfe, Jennifer J. Krauel, Devin N. Jones, Amy L. Russell, and Veronica A. Brown. 2018. Rapid Range Expansion of the Brazilian Free-Tailed Bat in the Southeastern United States, 2008–2016. *Journal of Mammalogy* 99(2):312–320.

Meyer, Christoph F. J., Matthew J. Struebig, and Michael R. Willig. 2015. Responses of Tropical Bats to Habitat Fragmentation, Logging, and Deforestation. In *Bats in the Anthropocene: Conservation of Bats in a Changing World*. Edited by Christian C. Voigt and Tigga Kingston. New York: Springer. 63–103.

Mickleburgh, Simon P., Anthony M. Hutson, and Paul A. Racey. 2002. A Review of the Global Conservation Status of Bats. *Oryx* 36(1):18–34.

Mildenstein, Tammy, Iroro Tanshi, and Paul A. Racey. 2015. Exploitation of Bats for Bushmeat and Medicine. In *Bats in the Anthropocene: Conservation of Bats in a Changing World*. Edited by Christian C. Voigt and Tigga Kingston. New York: Springer. 325–375.

Mohd-Azlan, Jayasilan, Joon Yee Yong, Nabila Norshuhadah Mohd Hazzrol, Philovenny Pengiran, Arianti Atong, and Sheema Abdul Aziz. 2022. Local Hunting Practices and Perceptions Regarding the Distribution and Ecological Role of the Large Flying Fox (Chiroptera: Pteropodidae: *Pteropus vampyrus*) in Western Sarawak, Malaysian Borneo. *Journal of Threatened Taxa* 14(1):20387–20399.

Monadjem, Ara, Magnus Ellstrom, Cristina Maldonaldo, and Nicolas Fasel. 2010. The Activity of an Insectivorous Bat *Neoromicia nanus* on Tracks in Logged and Unlogged Forest in Tropical Africa. *African Journal of Ecology* 48(4):1083–1091.

Monck-Whipp, Liv, Amanda E. Martin, Charles M. Francis, and Lenore Fahrig. 2018. Farmland Heterogeneity Benefits Bats in Agricultural Landscapes. *Agriculture, Ecosystems & Environment* 253(February):131–139.

Oedin, Malik, Fabrice Brescia, Alexandre Millon, Brett P. Murphy, Pauline Palmas, John C. Z. Woinarski, and Eric Vidal. 2021. Cats *Felis catus* as a Threat to Bats Worldwide: A Review of the Evidence. *Mammal Review* 51(3):323–337.

Pauwels, Julie, Isabelle Le Viol, Yves Bas, Nicolas Valet, and Christian Kerbiriou. 2021. Adapting Street Lighting to Limit Light Pollution's Impacts on Bats. *Global Ecology and Conservation* 28(August):e01648.

Pschonny, Sandra, Jan Leidinger, Rudolf Leitl, and Wolfgang W. Weisser. 2022. What Makes a Good Bat Box? How Box Occupancy Depends on Box Characteristics and Landscape-Level Variables. *Ecological Solutions and Evidence* 3(1):e12136.

Put, Julia E., Lenore Fahrig, and Greg W. Mitchell. 2019. Bats Respond Negatively to Increases in the Amount and Homogenization of Agricultural Land Cover. *Landscape Ecology* 34(8):1889–1903.

Rabie, Paul A., Brandi Welch-Acosta, Kristen Nasman, Susan Schumacher, Steve Schueller, and Jeffery Gruver. 2022. Efficacy and Cost of Acoustic-Informed and Wind Speed-Only Turbine Curtailment to Reduce Bat Fatalities at a Wind Energy Facility in Wisconsin. *PLoS ONE* 17(4):e0266500.

Romano, W. Brad, John R. Skalski, Richard L. Townsend, Kevin W. Kinzie, Karyn D. Coppinger, and Myron F. Miller. 2019. Evaluation of an Acoustic Deterrent to Reduce Bat Mortalities at an Illinois Wind Farm. *Wildlife Society Bulletin* 43(4):608–618.

Rowse, E. G., D. Lewanzik, E. L. Stone, S. Harris, and G. Jones. 2015. Dark Matters: The Effects of Artificial Lighting on Bats. In *Bats in the Anthropocene: Conservation of Bats in a Changing World*. Edited by Christian C. Voigt and Tigga Kingston. New York: Springer. 187–213.

Rueegger, Niels. 2016. Bat Boxes: A Review of Their Use and Application, Past, Present and Future. *Acta Chiropterologica* 18(1):279–299.

Rueegger, Niels, Ross L. Goldingay, Brad Law, and Leroy Gonsalves. 2019. Limited Use of Bat Boxes in a Rural Landscape: Implications for Offsetting the Clearing of Hollow-Bearing Trees. *Restoration Ecology* 27(4):901–911.

Russo, Danilo, and Leonardo Ancillotto. 2015. Sensitivity of Bats to Urbanization: A Review. *Mammalian Biology* 80:205–212.

Sandoval-Herrera, Natalia, Linda Lara-Jacobo, Juan S. Vargas Soto, Paul A. Faure, Denina Simmons, and Kenneth Welch. 2022. Common Insecticide Affects Spatial Navigation in Bats at Environmentally-Realistic Doses. *BioRxiv* doi. org/10.1101/2022.09.14.508021.

Sirami, Clélia, David Steve Jacobs, and Graeme S. Cumming. 2013. Artificial Wetlands and Surrounding Habitats Provide Important Foraging Habitat for Bats in Agricultural Landscapes in the Western Cape, South Africa. *Biological Conservation* 164(August):30–38.

Stahlschmidt, Peter, Achim Pätzold, Lisa Ressl, Ralf Schulz, and Carsten A. Brühl. 2012. Constructed Wetlands Support Bats in Agricultural Landscapes. *Basic and Applied Ecology* 13(2):196–203.

Stepanian, Phillip M., and Charlotte E. Wainwright. 2018. Ongoing Changes in Migration Phenology and Winter Residency at Bracken Bat Cave. *Global Change Biology* 24(7):3266–3275.

Stone, Emma Louise, Stephen Harris, and Gareth Jones. 2015. Impacts of Artificial Lighting on Bats: A Review of Challenges and Solutions. *Mammalian Biology* 80(3):213–219.

Streicker, Daniel G., Sergio Recuenco, William Valderrama, Jorge Gomez Benavides, Ivan Vargas, Víctor Pacheco, Rene E. Condori Condori, et al. 2012. Ecological and Anthropogenic Drivers of Rabies Exposure in Vampire Bats: Implications for Transmission and Control. *Proceedings of the Royal Society B: Biological Sciences* 279(1742):3384–3392.

Tanalgo, Krizler C., Hernani F. M. Oliveira, and Alice Catherine Hughes. 2022. Mapping Global Conservation Priorities and Habitat Vulnerabilities for Cave-Dwelling Bats in a Changing World. *Science of the Total Environment* 843(October):156909.

Tanalgo, Krizler Cejuela, Tuanjit Sritongchuay, Angelo Rellama Agduma, Kier Celestial Dela Cruz, and Alice C. Hughes. 2023. Are We Hunting Bats to Extinction? Worldwide Patterns of Hunting Risk in Bats Are Driven by Species Ecology and Regional Economics. *Biological Conservation* 279(March):109944.

Turner, Gregory G., Brent J. Sewall, Michael R. Scafini, Thomas M. Lilley, Daniel Bitz, and Joseph S. Johnson. 2022. Cooling of Bat Hibernacula to Mitigate White-Nose Syndrome. *Conservation Biology* 36(2):e13803.

Vanderwolf, Karen J., Lewis J. Campbell, Tony L. Goldberg, David S. Blehert, and Jeffrey M. Lorch. 2020. Skin Fungal Assemblages of Bats Vary Based on Susceptibility to White-Nose Syndrome. *The ISME Journal* 15(3):909–920.

Voigt, Christian C., Kendra L. Phelps, Luis F. Aguirre, M. Corrie Schoeman, Juliet Vanitharani, and Akbar Zubaid. 2015. Bats and Buildings: The Conservation of Synanthropic Bats. In *Bats in the Anthropocene: Conservation of Bats in a Changing World*. Edited by Christian C. Voigt and Tigga Kingston. New York: Springer. 427–462.

Weaver, Sara P., Cris D. Hein, Thomas R. Simpson, Jonah W. Evans, and Ivan Castro-Arellano. 2020. Ultrasonic Acoustic Deterrents Significantly Reduce Bat Fatalities at Wind Turbines. *Global Ecology and Conservation* 24:e01099.

Webala, Paul W., Jeremiah Mwaura, Joseph M. Mware, George G. Ndiritu, and Bruce D. Patterson. 2019. Effects of Habitat Fragmentation on the Bats of Kakamega Forest, Western Kenya. *Journal of Tropical Ecology* 35(6):260–269.

Welbergen, Justin A., Stefan M. Klose, Nicola Markus, and Peggy Eby. 2007. Climate Change and the Effects of Temperature Extremes on Australian Flying-Foxes. *Proceedings of the Royal Society B: Biological Sciences* 275(1633):419–425.

Welch, Jessica Nicole, and Jeremy M. Beaulieu. 2018. Predicting Extinction Risk for Data Deficient Bats. *Diversity* 10(3):63.

Welch, Jessica Nicole, and Christy Leppanen. 2017. The Threat of Invasive Species to Bats: A Review. *Mammal Review* 47(4):277–290.

Williams-Guillén, Kimberly, Elissa Olimpi, Bea Maas, Peter J. Taylor, and Raphaël Arlettaz. 2015. Bats in the Anthropogenic Matrix: Challenges and Opportunities for the Conservation of Chiroptera and Their Ecosystem Services in Agricultural Landscapes. In *Bats in the Anthropocene: Conservation of Bats in a Changing World*. Edited by Christian C. Voigt and Tigga Kingston. New York: Springer. 151–186.

Woods, Michael, Robbie A. McDonald, and Stephen Harris. 2003. Predation of Wildlife by Domestic Cats *Felis catus* in Great Britain. *Mammal Review* 33(2):174–188.

Zhelyazkova, V., A. Hubancheva, G. Radoslavov, N. Toshkova, and S. J. Puechmaille. 2020. Did You Wash Your Caving Suit? Cavers' Role in the Potential Spread of *Pseudogymnoascus destructans*, the Causative Agent of White-Nose Disease. *International Journal of Speleology* 49(2):149–159.

# Index

BILL BROKAW

# About the Author

Behavioral ecologist and science communicator **Alyson Brokaw**'s professional passion is communicating about the diversity and biology of bats, as well as combating myths that they are scary or dangerous. As she earned her PhD in Ecology and Evolutionary Biology from Texas A&M University, her affinity for these amazing creatures led her to study how they use their sense of smell during hunting and social communication. She has studied wildlife around the world, including Mexico, Belize, Panama, Argentina, and Tanzania, while also publishing in peer-reviewed journals multiple articles on bat behavior and other topics. She is a narrator for March Mammal Madness and an active science communicator on social media (where her handle is @alyb_batgirl). When not talking your ear off about bats, she can be found chasing down discs on the Ultimate Frisbee field, cuddling with her dog, or with her nose buried deep in a fantasy novel. Learn more about Alyson and her research on her website AlysonBrokaw.com.